B

Manfred N. Polrin
1. 12. 1988

Zum Geburtstag von
Jörg und Katja!

Basel, 30. 12. 88

Architektur
im Zusammenhang

Herausgegeben von
Rudolf Schilling

John McKean

Walter Segal's Life, Work and Influence

LEARNING FROM SEGAL
VON SEGAL LERNEN

Walter Segals Leben, Werk und Wirkung

„meine Hochachtung!"

Birkhäuser Verlag
Basel · Boston · Berlin

Deutsche Übersetzung
Olinde Riege, Weesen

Library of Congress Cataloging in Publication Data
McKean, John, 1943–
    Learning from Segal.

    Bibliography: p.
    Includes index.
    1. Segal, Walter, d. 1985. 2. Architects–Great Britain–Biography.
I. Title. II. Title: Von Segal lernen.
    NA997.S45M35 1989   720'.92'4   88-34228
    ISBN 3-7643-1999-2

CIP-Kurztitelaufnahme der Deutschen Bibliothek
McKean, John:
Learning from Segal : Walter Segal's life, work and influence = Von
Segal lernen / John Mckean. Übertr. ins Dt.: Olinde Riege. – Basel ;
Boston ; Berlin : Birkhäuser, 1989
    (Architektur im Zusammenhang)
    ISBN 3-7643-1999-2
NE: Segal, Walter [Ill.]

© 1989 Birkhäuser Verlag, Basel
All drawings by Walter Segal are © by John A. Segal; all rights
reserved. Credits see p. 221.
Buch- und Umschlagsgestaltung:
Justin Messmer, Basel
Printed in Germany
ISBN 3-7643-1999-2

Vorbemerkung zur deutschen Übersetzung
Der Originaltext des Buchs ist der englische. Bei der Übertragung
ins Deutsche wurden einige wenige Abschnitte weggelassen.
Ferner sind die Anmerkungen, die im englischen Text jeweils am
Ende jedes Kapitels stehen, bei der deutschen Fassung in den
Haupttext integriert. Die Quellenverweise zu den Zitaten aus Walter
Segals eigenen Texten stehen jeweils am Schluß jedes Kapitels der
englischen Fassung. Sie sind nach der Reihenfolge des Erschei-
nens im Text geordnet und beziehen sich auf die chronologische
Liste der Originaldokumente am Schluß des Buches (Anhang B).

Contents / Inhaltsverzeichnis

Preface

This text introduces a fascinating character, his *Welt-anschauung* and pertinent ideas for our times. Walter Segal was an inveterate and always fascinating designer. He never laid up his drawing pen during slack periods; he was a considerable writer and he left a life's archive. Therefore this volume must be an *hors d'oevre* – or as we say in English, »appetiser«, to loosen and activate the sensitivty of the taste buds, for what surely must follow.

I am grateful first to Walter Segal who, with Florian Adler as publisher, asked me to undertake a monograph on him a decade ago. Sadly that project ran out of backing and, now of course after Segal's death, this is a very different text. Mistakes and mysteries can no longer be easily clarified by him; nor can what reads now as clear be made paradoxical or delphic.

One of my central aims is to express Walter Segal's world as much as possible in his own words. I am therefore particularly grateful to Bob Wills and Cherie Yeo for the use of tapes they made of discussions with Walter Segal over two days in 1984.

This book wouldn't have been possible without the enthusiasm of Rudolf Schilling as the project's midwife. I have received encouragement and assistance from such a varied host of Segal's friends that their assembly together would make an astonishing but most convivial party; I gratefully name them below. My thanks go especially to Moran Segal, not least for helping me live with a presbyterian conscience, and Marion Adler for putting up with it.

During this time, my son's cat Wally has died of old age. She was lovable but always neurotic; we say it was because her eponymous donor used to pick her up by the tail when a tiny kitten, and throw her about. So he told us, with a naughty twinkle in his eye. I guess I believed him.

John McKean, London, May 1988

Acknowledgements

For sharing their memories and thoughts of Walter Segal, my thanks to John Allsopp, Ken Atkins, Alan Blanc, Peter Blundell Jones, Colin Boyne, John Brandon-Jones, Jon Broome, Philip Christou, Neville Conder, Tony Cooper, Charlotte Ellis, Otto Koenigsberger, Monica Pidgeon, Julius Posener, Sir Philip Powell, Brian Richardson, John Segal, Peter Stead, Pete Sutton, Sir Roger Walters, Colin Ward.

While writing this book, I was involved on the periphery of an exhibition, the brainchild of Florian Beigel, which celebrates Walter Segal. Ideas were developed during this time, my thoughts clarified and my material knowledge increased in discussions with Peter Rich, John Rae and Bob Wills, Florian Beigel and especially Philip Christou.

All drawings by Walter Segal are from his archive which remains the copyright of John Segal and is shortly to be deposited on loan with the Drawings Collection of the Royal Institute of British Architects.

I have also been assisted by Dr Marianne Eaton-Krauss, the Egyptologist to whom Segal's work in Egypt was entrusted, and Greta Stroech of Fondation Arp. Margot Adler helped greatly with translation of German texts. Tricia Noble and her colleagues at the PNL library kindly helped obtain obscure published material. Shirley McPherson, photographic Librarian at the Architectural Press was most helpful with illustrations.

None of those mentioned, of course, is responsible for opinions expressed or errors contained herein. And I must apologise to all those who helped me with fascinating material for which there just is not room in such a modest volume.

Vorwort

Dieser Text stellt einen faszinierenden Charakter vor, eine Weltanschauung und Ideen, die in unsere Zeit fortwirken. Walter Segal war ein unermüdlicher und eindrucksvoller Zeichner; niemals legte er seinen Zeichenstift aus der Hand. Und er war ein beachtlicher Schriftsteller, wenn auch einige seiner fundiertesten Schriften kaum verbreitet wurden. Da er alles peinlich genau sammelte und aufbewahrte, hinterließ er ein komplettes Archiv seines Lebenswerks. Deshalb sollte dieser Band auch als ein hors d'oeuvre, als Appetithappen, betrachtet werden, der die Geschmacksnerven für etwas anregt, das noch folgen sollte.

Dankbar bin ich in erster Linie Walter Segal, der mich schon vor einem Jahrzehnt aufforderte, zusammen mit Florian Adler als Herausgeber eine Monografie über ihn in Angriff zu nehmen. Leider erhielt das Projekt damals nicht die notwendige Unterstützung. Nun, nach Segals Tod, ist dieses natürlich ein ganz anderer Text. Fehler und Unklarheiten können nicht mehr einfach von ihm selbst korrigiert werden.

Es ist mir ein zentrales Anliegen, die Welt um Walter Segal so weit wie möglich in seinen eigenen Worten auszudrücken. Ich danke deshalb ganz besonders Bob Wills und Cherie Yeo dafür, daß sie mir gestatteten, die Tonbänder einer zweitägigen Diskussion mit Walter Segal, im Jahre 1984, zu benutzen.

Dieses Buch wäre nicht möglich gewesen ohne die Begeisterung Rudolf Schillings, gewissermaßen der Hebamme des Projekts, und die Unterstützung durch John Segal, dem Sohn des Architekten. Ermutigung und Unterstützung wurde mir durch eine vielfältige Schar von Segals Freunden zuteil; es würde eine erstaunliche, aber sehr fröhliche Party geben, könnte man sie alle zusammenbringen. Ganz besonders gehört mein Dank Moran Segal, nicht zuletzt, weil sie mir half, mit einem presbyterianischen Gewissen zu leben, und Marion Adler, weil sie das zu tragen wußte.

John McKean, London, Mai 1988

Dank

Ich danke John Allsopp, Ken Atkins, Alan Blanc, Peter Blundell Jones, Colin Boyne, John Brandon-Jones, Jon Broome, Philip Christou, Neville Conder, Charlotte Ellis, Otto Königsberger, Monica Pidgeon, Julius Posener, Sir Philip Powell, Brian Richardson, Peter Stead, Pete Sutton, Sir Roger Walters, Colin Ward dafür, daß sie ihre Erinnerungen an Walter Segal mit mir teilten.

Während ich dieses Buch schrieb, war ich am Rande in die Vorbereitung einer Ausstellung zu Ehren Walter Segals involviert. Ich konnte während dieser Zeit meine Materialkenntnisse erweitern in Diskussionen mit Peter Rich, John Rae, Bob Wills, Florian Beigel und ganz besonders mit Philip Christou. Die Ausstellung wurde vom 17. Mai bis 12. Juni 1988 in der Royal Festival Hall, London, erstmals gezeigt.

Alle Zeichnungen von Walter Segal in diesem Buch stammen aus dem Archiv, dessen Copyright bei John Segal bleibt, der es demnächst als Leihgabe der Grafischen Sammlung des Royal Institute of British Architects übergeben wird.

Weiter wurde ich unterstützt durch Dr. Marianne Eaton-Krauss, der Ägyptologin, der Segals ägyptische Aufzeichnungen anvertraut sind, und Greta Stroech von der Fondation Arp. Margot Adler war sehr hilfreich bei der Übersetzung der deutschen Texte. Tricia Noble und ihre Kollegen von der Bibliothek des Polytechnic of North London halfen mir, schwer zugängliches Material aufzutreiben.

Selbstverständlich ist keine und keiner der hier Erwähnten verantwortlich für geäußerte Meinungen oder für Irrtümer. Außerdem möchte ich alle jene um Verzeihung bitten, die mir mit Material aushalfen, das dann leider in diesem Band keinen Platz mehr fand.

Walter Segal 1974.

Phaedrus talks to Socrates of a conversation with his friend Eupalinos, an architect:

Phaedrus

»Phaedrus«, he was saying to me, »the more I meditate on my art, the more I practice it; the more I think and act, the more I suffer and rejoice as an architect; and the more I feel my own being with an ever surer delight and clarity.

»I go astray in my long spells of waiting; I find myself again by the surprises I give myself; by means of these successive degrees of my silence, I advance in my own edification; and I draw near to such an exact correspondence between my aims and my powers, that I seem to myself to have made of the existence that was given me a sort of human handiwork.

»By dint of constructing,« he put it with a smile, »I truly believe that I have constructed myself.«

Socrates

To construct oneself, to know oneself – are these two, distinct acts or not?

Phaedrus

... and he added: »I have sought accuracy in my thoughts, so that being engendered by the consideration of things they might be changed as though of their own accord into the acts of my art I have apportioned my attentions; I have arranged the problems in another order; I begin where I finished off formerly, so as to go a little further... I am niggardly of musings, I conceive as though I were executing. What I think is feasible, and what I do is related to the intelligible.... And then... listen, Phaedrus,« he went on to say, »this little temple which I built for Hermes, a few steps from here, if you could know what it meant to me! There where the passer-by sees but an elegant chapel – tis but a trifle: four columns, a very simple style – there I have enshrined the memory of a bright day in my life.«

Paul Valéry, Eupalinos or The Architect

Phaidros sprach zu Sokrates:

Phaidros

Eines Tages, lieber Sokrates, sprach ich von eben diesen Dingen mit meinem Freund Eupalinos.
– Phaidros, sagte er mir, je mehr ich über meine Kunst nachdenke, desto mehr übe ich sie aus; je mehr ich denke und handle, desto mehr leide und freue ich mich in meiner Eigenschaft als Architekt; – und um so mehr fühle ich mich selbst mit einer Wollust und einer Klarheit, die immer noch an Sicherheit gewinnen.
Ich verirre mich in lange Wartezeiten; ich finde mich wieder in Überraschungen, die ich mir selbst bereite; und mittels dieser allmählichen Abstufungen in meinem Schweigen schreite ich in meiner eigenen Erbauung vor; ich nähere mich einer so genauen Beziehung zwischen meinen Wünschen und meinen Fähigkeiten, daß es mir scheint, als hätte ich aus der Existenz, die mir gegeben wurde, ein Menschenwerk gemacht.

Indem ich baute, warf er lächelnd hin, habe ich mich, glaube ich, selbst erbaut.

Sokrates

Sich erbauen und sich selbst erkennen, sind das zwei getrennte Akte oder nicht?

Phaidros

... und er fügte hinzu: Ich habe das Richtige gesucht in den Gedanken, damit sie, in klarer Weise hervorgegangen aus der Betrachtung der Dinge, sich wie von selbst verwandeln in die Handlungen meiner Kunst. Ich habe meine Aufmerksamkeiten verteilt; ich habe die Probleme umgeordnet; ich fange an, wo ich früher aufgehört habe, um ein bißchen weiterzugehen ... Ich bin geizig im Träumen. Wenn ich mir etwas vorstelle, ist es schon immer, als führte ich etwas aus. Aber das, was ich denke, läßt sich ausführen; und das, was ich ausführe, geht auf ein Einsehen zurück ... Und dann ... Höre, Phaidros (sagte er mir noch), der kleine Tempel, den ich einige Schritte von hier für Hermes gebaut habe, wenn du wüßtest, was er für mich bedeutet! – Wo der Vorübergehende nichts sieht als eine elegante Kapelle – eine Kleinigkeit: vier Säulen in sehr einfachem Stil –, da habe ich die Erinnerung an einen lichten Tag meines Lebens untergebracht.

Paul Valéry, Eupalinos

Guiding Principles

Walter Segal (1907–1985) was an architect who grew up in the centre of European Modernism between the World Wars. From there he developed, steadily and quietly, in a comprehensible and coherent direction. But in never losing his unblinking common-sense, passionate scepticism and openness to the potential around him, he trod, for more than half a century as an architect, a singular path. Indeed, in his last fifteen years, at an age – from mid sixties to nearly eighty – when most of us would have been living on our past, his architectural achievement flowered with new strength. While he was always fascinated by the new, and his work developed continually, there is also a wholeness and consistency about Segal's life which is remarkable. This quality lets Segal offer us today a unique bridge between the aims of that great moment of Modernism and the world of the late 1980s.

On the one side of this bridge, we see the culture of Germanic modern art: where Segal's father was an important painter; where family friends of his youth included Tzara, Arp and Oud; Mies van der Rohe, Klee and Kandinsky; and where, a few years later, architects Gropius, Mendelsohn and particularly his friend Bruno Taut each at different times wanted Walter Segal to work with them.

Then, on the other side, we see the culture of late twentieth-century housing need: where Segal's clients, arriving randomly from Local Authority lists of those needing housing, became friends and allies, building their homes and transforming their lives in the 1980s.

As the twentieth century closes, we struggle to relocate issues of dwelling and place-making at the centre of our concerns. We see dwelling as an act rather than an object, housing as a verb and the housing architect as enabler rather than provider, as facilitating this deep human need. Segal developed a way of making architecture which lets this flourish: which allows the unhoused and unskilled simply to build their own homes, not in the »third world« but in the cradle of the »industrial revolution« itself.

The path Walter Segal drew between these two worlds, between the 1920s and the 1980s, shows us forgotten links. It can illuminate in startlingly new ways the real strengths of that »modernism« of which he never lost sight. It has, of course, nothing to do with stylistic formalism or of deference to the memory of great masters.

In one sense Segal's career illustrates that phrase in which Lewis Mumford summed up »modernism«: »The age of the crustacean building has given way to the age of vertibrates.« In a period of fancy shells and, perhaps, less backbone, it reminds us that these issues are deeper than those of fashion.

In another sense his career – which focused on housing – illustrates the essential modern theme: »equivalence of opportunity«, the provision of equal quality but not the same housing environment for all. This informed Segal whether planning rooms for the individuals within a house, locating the different houses within a neighbourhood, or enabling different people, without capital resources, to create their own homes. This »equivalence of opportunity« was the unspoken concern of the new architecture he studied as a youth in Holland, Weimar Germany and beyond; it was the focus of his own wider studies of indigenous housing in the 1930s; it concentrated his mind into a virtuosic collection of radical house plans and layouts in the 1940s. Some of these he built in the 1950s and its principles he reassessed, building radically different houses in the 1960s. This was a way of making architecture whose potential he re-examined in the 1970s, finally turning the machine for living into a language for assembling humane dwellings.

So Segal's path throughout was marked out by a lifelong attachment to a few guiding principles. The goals sought the non-hierarchical, equality without sameness, and social conviviality. The method was a belief in reasoning rather than imagery, and a lack of preconceptions.

The power of reasoning

At the centre is the consistent thread of faith in the power

Leitlinien

Der Architekt Walter Segal (1907–1985) wuchs zwischen den beiden Weltkriegen im geistigen Zentrum der europäischen Moderne auf. Aus dieser Mitte entwickelte er sich stetig in einer klaren und logischen Richtung. Ohne jemals seinen unbeirrbaren gesunden Menschenverstand, seinen leidenschaftlichen Skeptizismus und seine Offenheit zu verlieren, ging er über mehr als ein halbes Jahrhundert seinen eigenen Weg. Auf diese Weise schlägt Segals Leben und Werk eine Brücke zwischen jener großen Zeit der Moderne und dem Ende des Jahrhunderts.

Segals Vater war ein bedeutender Kunstmaler. Zu den Freunden der Familie zählten Tzara, Arp, Schwitters, Klee, Kandinsky und Mies van der Rohe. Zu verschiedenen Zeitpunkten bemühten sich Gropius, Mendelsohn und Bruno Taut um Walter Segal als Mitarbeiter. Das ist der eine Brückenkopf.

Auf der anderen Seite treffen wir einen Architekten an, der sich mit den Wohnbedürfnissen des auslaufenden 20. Jahrhunderts befaßt, mit Wohnungssuchenden, die aus Listen von Beamten ausgelost werden. Diese Wohnungssuchenden ohne Privilegien und ohne Geld in der Tasche sind seine Klienten, Freunde und Verbündeten. Mit ihnen baut er. Mit ihnen definiert er die Wohnung als einen Vorgang, nicht als Objekt. Der Architekt stellt keine fertigen Häuser zur Verfügung, sondern versteht Behausen als eine Tätigkeit, die es den Unbehausten und Unausgebildeten erlaubt, sich ihre eigenen Häuser zu bauen.

Der Weg, den Walter Segal von der Welt der Zwanziger in die Welt der Achtziger Jahre gegangen ist, zeigt vergessene Verbindungen und wirft ein verblüffendes Licht auf die wahren Stärken der »Moderne«, die er niemals aus den Augen verlor. Das hat weder mit Form noch Stil noch Ehrfurcht vor den großen Meistern etwas zu tun. Segals Laufbahn läßt sich eher mit Lewis Mumfords Ausspruch illustrieren: »Das Zeitalter des Bauens in der Art der Krustentiere ist vorbei; es hat den Weg freigegeben für das Zeitalter der Wirbeltiere.« In einer Zeit, da verspielte Gehäuse beliebt sind, befaßt sich Segal mit den Prinzipien des Behausens. In seiner Tä-

tigkeit, die ganz auf den Wohnungsbau ausgerichtet war, ging es stets um das Thema »Chancengleichheit«, nämlich um die Bereitstellung guten Wohnraums für alle. Ob er Individuen half, die innere Organisation ihres Hauses zu planen, oder ob er Häuser in einer Nachbarschaft anordnete, oder ob er es Leuten ohne Eigenkapital ermöglichte, sich selber Häuser zu bauen: Stets war »Chancengleichheit« das unausgesprochene Anliegen, das er als junger Mann in Holland und im Deutschland der Weimarer Republik studiert und in sich aufgesogen hatte. Er verstand »Chancengleichheit« als das Ziel des neuen Bauens der Dreißiger Jahre. Es animierte ihn, ein System zu entwickeln, das schließlich zu radikal anderen Baumethoden führte als den gewohnten – und zu radikal anderen Häusern.

Die Macht der Vernunft

Walter Segal glaubte an die Macht der Vernunft. Sein gesunder Menschenverstand war ungewöhnlich gut entwickelt. Er sah alle Dinge, als würden sie zum Erstenmal gesehen, und formulierte sie, als würden sie zum Erstenmal formuliert. Darum zeichnen sich alle seine Arbeiten und Entwürfe durch eine »ungewöhnliche Gewöhnlichkeit« aus, von der Methode, wie Einzelteile zu einem Bau zusammengesetzt werden, bis zur Realisierung von sozialer Architektur.

Seit seiner Kindheit gab es bei Walter Segal keinen Raum für Bluff. Immer waren seine Konstruktionen genau berechnet, egal ob aus Backstein, Holz oder Metall. Beton vermied er; er hielt ihn für »ungenau«; sein Verhalten war ihm unverständlich. Andererseits betrachtete er das Angebot eines Baustoffhändlers als etwas, das man ohne vorgefaßte Meinung betrachten, auf mögliche Anwendungen »durchschauen« solle.

Er wollte die Materialien verstehen und die Art, wie sie funktionieren, wenn man sie zusammensetzt. Er setzte sein Vertrauen dabei auf einfache Prinzipien, selbst wenn die Lösung höchst unüblich schien. Daß er sich entschloß, 50mm Wasser als Deckmaterial auf das Flachdach seines Hauses zu bringen, ausgerechnet in einem Moment, als das Versagen der Flachdächer zu

A client-built Segal frame before being clad.

Im Selbstbau erstelltes Fachwerk nach der Segal-Methode vor der Auskleidung.

of reasoning. The first characteristic is an uncommon common-sense. Seen at all scales from how the small pieces of building are put together, right up to how a social architecture is formed, this is marked by an extraordinary ordinariness.

One side of this theme demands a sensitivity to the material world, and an openness to its possibilities. Seeing things as they are, as if for the first time; seeing their possibility beyond the given context, the given culture. That is what stretches culture and this quality places Segal firmly in an intellectual tradition. (Such quick-wittedness, a transformational awareness with a twinkle in the eye, of course is not alien to the sensibility of Dada.)

The theme's other face demands a sensitivity to the human world, and an openness to its potential. A faith in observation of human action and rational research into how we perceive spaces and occupy buildings; sensitivity to the nuances of behaviour and thus to understanding what makes places calm, convivial, pleasant to be in; and an openness to the creative roles of anyone in the making of places in which we can dwell.

Of these two facets, the first is based in understanding the sensate world of things – and an insatiable curiosity not just as to how things work but to how they *might* work. From childhood, there was no room with Walter Segal for bluffing. Whether in brick, metal or timber, his structures were precisely calculated. (Concrete he avoided, because of its imprecision; its behaviour was incomprehensible.) Conversely, he treated a builders merchant stock as components which, if looked at without preconception, reveal new possible uses.

He had to understand his materials and how they operated when put together; putting his trust in simple principles, even when the solution seemed most unusual. That he decided to use 50mm of water as the finishing material on his domestic flat roof, despite the fact that current architectural news concerned widespread flat-roof failure, was sheer commonsense (moreover, its weight held the building down, it provided insulation and immediately showed up any structural movement or subsidence). It wasn't cussedness; but naturally, he enjoyed being so clearly against the grain.

The second of these facets is based in understanding the world of human behaviour, of history and anthropology – an insatiable desire to understand how we tick. Through his life Segal observed how we inhabit dwellings, he read voraciously, but he trusted his senses most of all. He produced studies of peasant housing in Egypt and in the Baleares; and more than anything else he designed inhabitable homes.

In the social world, there was no room for bluffing either. He had spent his childhood amongst the most loquatious painters, »and to this day I am convinced they did not understand what each other said. But it did not seem to matter, and when it lead to sound poems all was well!«[1]

Certainly, architecture is not just the fulfilment of direct utliitarian requirements but, equally, it is not a business of visual images. To Segal, architecture must appeal to the mind, transcending utility and imagery.

Travelling light : »the luggage problem«
Walter Segal's guiding principles can also be characterised by a remarkable Zen-like ability to reject the inessential. The jettisoned material ranged from cultural preconceptions to wasteful modes of architectural practice and building construction. There was a deep theme of economy of effort, controlled by reason.

As he memorably put it: »Architecture had already arrived at a station where another train was waiting to take it into a new world; but it hesitated to change, and when it did it took more of its luggage than it needed. So a luggage problem remains. And now, oh dear,there is a new train waiting. Shall we ever learn to travel light?« His career was a continuous process of freeing himself from one »tyrrany« (his word) after another. Walter Segal described himself as »a fierce individualist«; he liberated himself from holy cows and received opinions from »heroes« and from »superstitions« about architecture. He liberated himself from unproductive tasks: he drew simply, increasingly freehand (perhaps over squared paper) and on small (A4) sheets, free of drawing boards and instruments, free from reliance on dyeline printers for reproducing drawings.

Segal's way of designing freed him from quantity surveyors and from engineers. His developed form of professional practice freed him from delegating down to architectural assistants or up to general building contractors. Working this way and without employees, he was free of office management; working where possible

Verandah at Lewisham 1980.
Veranda in Lewisham 1980.

**»Garden view of row of castellated terrace houses.«
Designed and drawn by Segal in 1943.
»Gartenseite einer Zeile verdichteter Reihenhäuser.«
Entworfen und gezeichnet von Segal 1943.**

directly as »builder« and subcontracting tradeswork, he was free of job management as understood by *architects*. His preferred method of building freed him from wet trades (no concrete, no bricks and mortar, no plaster).

Perhaps most of all he freed himself from the past. »The tradition of all the dead generations weighs like a nightmare on the brain of the living,« said Marx[2]. Segal, who with mercurial slipperiness managed to avoid this burden, recognised it only too literally in architecture: »The heritage of the past is often a formidable obstacle to the realisation of modern solutions,« he said.

As a contrasting image, all his life Walter Segal kept a small, yellowed cutting from a German newspaper of the 1920s showing a spare, US steel frame under construction. And then, from the mid-1960s, Segal developed his own elegant method of building houses with frames and infillings which struck an immediate chord, becoming in itself an icon for others.

Economy of effort means that small is beautiful. It also means that effort is chanelled to where it is most effective. In the world of architecture, that means concentrating on the cerebral activity of designing. And in construction, it is in using the highest value material, calculated with precision to avoid waste, where it is most appropriate. Segal's expensive but minimal-section »stress-graded« timber frames exemplify this.

Himself, Segal somehow avoided being weighted down by »things«. For a journey, he carried a small knapsack and never a suitcase; as his wife said, he didn't *own* anything (except books). He was as free from neckties and coats as from humbug and styles.

But there were as many freedoms *for* as freedoms *from*. The freedom of designing within a set of rules increasingly allows more of the design work to be productive, and more of the architect's time to be engaged in designing. He liberated his individual clients through their engagement in the process of designing; and, finally, increasingly through the last decade, the freedom to make their own dwellings, freedom for »self-building« (which is more than just body-building).

Lifelong critique

The 1960s in Britain was filled with overweening housing projects, their soulless estates made up of vast blocks which were impossible to colonise and inhabit, and often – as we now all know to our cost – shoddily built and impossible to care for. Walter Segal's distaste for this world was not the judgement of hindsight. The carelessness which allowed seduction by images – the romance of »streets-in-the-air«, of municipal grandeur or whatever – with so little concern for detailed, human consequences; no less than the carelessness which allowed untried construction methods to be employed with so little supervisory control or concern for their detailed, material consequences; to Segal each showed unpardonable lack of true professional behaviour.

His own argument for humane mass housing, at high density but near the ground, and his brilliant designs to show the real feasibility of this aim, were published in the 1940s. If such ideas had been embraced by the providers of large-scale housing, the pattern of post-war Britain could have been very different.

Walter Segal, however, never had time for bemoaning or nostalgic observations. Quietly cultivating his

den wichtigsten Gesprächsthemen unter Architekten gehörte, war auch nur gesunder Menschenverstand. Das Gewicht des Wassers hielt das Gebäude am Boden, bot zusätzliche Isolierung und zeigte sofort jede Bewegung und Setzung der Konstruktion an. Es war keine Widerborstigkeit, aber er genoß es natürlich, so eindeutig gegen den Strich zu operieren. Das andere, das er verstehen wollte, war das menschliche Verhalten, die Geschichte der Anthropologie. Er wollte wissen, »wie wir funktionieren«. Sein ganzes Leben hindurch beobachtete Segal, wie wir in einer Wohung leben; er war unersättlich bei seiner Lektüre; aber in erster Linie vertraute er seinen Augen. Er verfaßte Studien über die Häuser der Bauern in Ägypten und auf den Balearen; und mehr als alles andere beschäftigte ihn das Entwerfen bewohnbarerer Häuser.

Auch in seiner sozialen Welt gab es keinen Raum für den Bluff. Er hatte seine Kindheit zwischen geschwätzigen Malern verbracht, »und bis heute bin ich überzeugt, daß sie niemals verstanden, was der andere sagte; aber das schien keine Rolle zu spielen, und wenn es zu guten Gedichten führte, war alles gut«.

Leicht reisen – »Das Gepäck-Problem«

Zu Walter Segals wesentlichsten Prinzipien gehörte eine bemerkenswerte Fähigkeit, alles Unwesentliche wegzulassen. Zum Unwesentlichen zählte er vorgefaßte Meinungen ebenso wie verschwenderische Methoden in der Praxis der Architekten. Es ging ihm prinzipiell um Ökonomie, kontrolliert durch den Verstand.

»Architektur war bereits an einer Endstation angekommen, und ein anderer Zug wartete, sie in eine neue Welt zu befördern; aber sie zögerten noch, umzusteigen, und als es endlich soweit war, wurde mehr Gepäck mitgeschleppt als nötig. So ist es beim Gepäck-Problem geblieben, und nun, Oh Gott, wartet schon ein neuer Zug. Sollen wir denn niemals lernen, leicht zu reisen?«

Seine ganze Karriere war ein Kampf gegen »Tyranneien«. Walter Segal bezeichnete sich selbst als einen »grimmigen Individualisten«; er befreite sich von heiligen Kühen und überlieferten Meinungen, von »Helden« und »Aberglauben« in der Architektur. Er befreite sich auch von unproduktiven Arbeiten: Er zeichnete sehr einfach, mehr und mehr freihändig (vielleicht über einem linierten Papier) und auf kleinen Bögen (A4), befreit vom Zeichenbrett und von Geräten, unabhängig von Lichtpausanlagen.

Damit wurde er auch unabhängig von Statikern, Bautechnikern und Ausschreibungsspezialisten. Seine Art, den Architektenberuf zu betreiben, befreite ihn vom Delegieren, sei es nach unten an die Assistenten, sei es nach oben an den Generalunternehmer. Ohne Angestellte war er frei vom Büro-Management. Indem er selbst, wenn immer möglich, als »Baumeister« fungierte und Handwerkerleistungen selber im Subkontrakt vergab, war er frei von der Baufführer-Aufgabe, wie sie in Architekturbüros verstanden wird. Auch gab es bei seiner Methode keine nassen Arbeiten (keinen Beton, keinen Mörtel, keinen Putz).

Vor allem jedoch befreite er sich selbst von der Vergangenheit. »Die Tradition all der toten Generationen lastet wie ein Alb auf den Gehirnen der Lebenden«, sagte Marx (in »The Eighteenth Brumaire of Louis Bonaparte«). Segal, der mit Merkur-gleicher Behendigkeit sich dieser Bürde entzog, hatte sie nur zu deutlich in der Architektur wahrgenommen. »Das Erbe der Vergangenheit ist oft ein schreckliches Hindernis bei der Verwirklichung moderner Lösungen«, sagte er.

Ein ganz anderes Bild: Sein ganzes Leben lang hob Segal einen vergilbten Zeitungsausschnitt aus den 20er Jahren auf, der einen fein dimensionierten amerikanischen Stahlrahmen während der Montage zeigt. Und dann, ab Mitte der 60er Jahre, entwickelte Segal seine eigene elegante Methode, Häuser mit Rahmen und Füllungen zu bauen, eine Methode, die sofort ein Vorbild für andere wurde.

Ökonomie in allen Bemühungen heißt: Small is beautiful. Ökonomie heißt auch, daß jede Bemühung dort eingesetzt werden sollte, wo sie besonders effektiv wird. In der Architektur heißt das, daß man sich auf die geistige Komponente des Entwerfens konzentrieren soll. Für die Konstruktion heißt das, daß man möglichst hochwertiges Material dort einsetzen soll, wo es am zweckmäßigsten ist, und daß man alles im voraus genau berechnet, um Verschwendung zu vermeiden. Segals Holzkonstruktionen, aus kostspieligsten, höchste Belastungen ertragenden Qualitäten, jedoch minimal dimensioniert, sind dafür beispielhaft.

Auch persönlich vermied es Segal, sich durch »Dinge« zu Boden drücken zu lassen. Für eine Reise benutzte er immer einen kleinen Rucksack; er trug niemals einen Koffer, und seine Frau sagte einmal, er besäße ja auch nichts, außer Büchern. Er war so frei von Krawatten und Mänteln, wie er frei war von Humbug und Stilen.

own small garden, he achieved a professional life which became a remarkable exemplar. His personal success was in standing alone, as a one-man architectural practice, ultimately being responsible only to craftsmen and clients with whom he had reached a close personal relationship.

Segal's radical approach to technical problems and his opening of the planning decisions and then building process to the users are strands of the one aim: cost-efficient, simple building which does not alienate the builder and which is a pleasure to inhabit. This unique professionalism, which appeals with its archetypal directness, stirs a profound response. Designing things that people get pleasure in making and, once made, people get pleasure in using (to paraphrase William Morris), is in a sense a modest aim. Yet their achievement in architecture, in Europe at the end of the twentieth century, is a massive and rare achievement.

The true professional method

But it is a vulgar misnomer to describe a concentration on the real issues of dwelling, of enabling the generation of ordinary and convivial homes, as »simple«. These are issues which outlast intellectual posing and competitions of connoiseurship. They stand against the primacy of the image. They also make for a more integrated sense of building, and a less alienated sense among the participants involved, be they architect or inhabitant, builder or craftsman.

Surely, says the aspiring young architect, there are some rules, some definite prescriptions for this honest architecture?

There *are* practical rules of procedure suitable to construction, comes the reply. But of course materials and means of building vary, so even these rules must be modified with such variation. However, »in architecture there is a *method of thinking* to be followed in all cases that present themselves, but there are no definite prescriptions or rules to be followed. This method is none other than the application of your reasoning faculty to all particular cases.«

That comment, in which we so strongly hear the echo of Segal, in fact predated him by a century. The speaker is »le grand cousin« Eugène, the architect in E E Viollet-Le-Duc's essay *How to Build a House* (»L'Histoire d'Une Maison«). In not offering the crutch for which so many clamour, his reply gave the key to

reading that fascinating book of Viollet's as a discourse on method: the method of common-sense, rooted in everyday reality. In that book we were shown a man fully conversant with all aspects and stages of construction: material, technical, aesthetic, legal and economic. His power was exercised through direct knowledge and contact. He was a man of the site, its material possibilities and processes. He drew not for the sake of drawing, but only to explain ideas which he had already worked out.

Walter Segal would have recognised a fellow professional in Eugène here; indeed it could be a description of Segal himself.

Viollet's book of 1875, ostensibly for teenagers, was called »the best book anyone could put in the hands of a young man thinking of a career in architecture.«[3] A century later, Jacques Aillagon[4] called it »one of the most moving pleas ever made for the profession of architect.«

Segal's career might equally be seen as a most moving plea for the possibility of the authentic professional architect today.

Notes and Quotations

1. To simplify referencing, Walter Segal quotations in each chapter are listed at the end of the footnotes.
2. Karl Marx, *The Eighteenth Brumaire of Louis Bonoparte.*
3. *Gazette des Beaux Arts*, Paris, 1880 (Viollet's hero »Eugène« in the English translation is unnamed, simply »le grand cousin« in the original French.)
4. *Architectural Design*, London, 1980

Quoting Walter Segal
To minimize textual interruption, quotations are noted at the end of each section, in the order of their appearance in the text. Segal's written English was very carefully crafted. His spoken English never lost its idiosyncratic European inflection, in intonation and construction. While inevitably intermingling these, I have tried *not* to iron them out.
Inevitably, almost all the quotations are from sources listed in *Appendix B* (selected texts by Segal). These are referenced by year, (B) or (C), and number in the list. The one other major and invaluable source is about eight hours of tape, recorded over the course of two days which Segal spent with Newcastle students, Cherie Yeo and Bob Wills in 1984. These are unpublished.

Quotations from Walter Segal in this section from: 1977(B)2; 1973(B)2; 1948(B).

Aber er war ebenso frei *für* etwas, wie frei *von* etwas. Die Freiheit des Entwerfens innerhalb fester Regeln erlaubt eine Steigerung der Produktivität, und der Architekt kann einen größeren Teil seiner Zeit auf den Entwurf verwenden. Er befreite auch seine Klienten, indem er sie am Entwurfsprozeß beteiligte. Mehr und mehr gab er ihnen auch die Freiheit, das eigene Haus selbst zu schaffen, selbst zu bauen.

Ein Leben in der Opposition

Wie andere Länder wurde England in den Sechziger Jahren mehr und mehr mit groben Wohnbauten überstellt. Die seelenlosen Siedlungen aus großen Blocks waren nicht in die Städte zu integrieren, und sie waren schlecht bewohnbar. Oft, wie wir zu unserem Schaden erfahren mußten, waren sie schäbig gebaut und sind sie schlecht instandzuhalten. Segals Abscheu vor dieser Un-Architektur wuchs nicht aus nachträglicher Einsicht. Seine Konzepte für Siedlungen hoher Wohndichte, jedoch nah am Boden, wurden schon in den Vierziger Jahren veröffentlicht. Es waren brilliante Entwürfe, die bewiesen, daß das Ziel, menschliche Siedlungen für die Massen zu realisieren, erreichbar ist. Hätten die Wohnbauproduzenten etwas von diesen Ideen aufgenommen, wäre das Bild der Nachkriegs-Städte ein anderes.

Segal nahm sich jedoch niemals Zeit für Klagen oder wehmütige Betrachtungen. Ruhig kultivierte er seinen eigenen kleine Garten und gestaltete sein berufliches Wirken auf seine Art. Sein persönlicher Erfolg bestand darin, daß er mit einem Ein-Mann-Büro enge persönliche Beziehungen zu schaffen wußte zu Handwerkern und Bauherren und nur ihnen gegenüber verantwortlich war.

Segals Einstellungen zu den Planungs- und Bauprozessen, sein Umgang mit künftigen Bewohnern sind auf ein einziges Ziel ausgerichtet: preiswerte, einfache Bauten zu schaffen, die zu erzeugen wie zu bewohnen eine Freude ist. Der Wunsch, Dinge zu entwerfen, die man gerne herstellt und mit Spaß benutzt, wirkt als bescheidener Anspruch. Diese Art von Berufseinstellung ist doch selbstverständlich. Und dennoch scheint es im Europa des ausgehenden 20. Jahrhunderts unglaublich schwer, diese einfachen Ziele zu erreichen.

Die wahrhaft professionelle Methode

Es ist ein landläufiger Irrtum, das eigentliche Ziel des Wohnungsbaues, nämlich den Bau gewöhnlicher und angenehmer Behausungen, als »simpel« abzutun. Es geht um etwas, das intellektuelle Selbstdarstellung und brilliante Lösungen überdauert. Es steht im Gegensatz zum Primat des Bildhaften. Es steht für eine integrierte Vorstellung vom Bauen: weniger Entfremdung zwischen den Beteiligten, Architekten und Bewohnern, Baumeistern und Handwerkern.

»Sicherlich gibt es dann«, sagt der ehrgeizige junge Architekt, »gewisse Regeln, einige definitive Rezepte für diese ehrenhafte Architektur?«

»Es gibt praktische Verfahrensregeln für das Bauen, ist die Antwort, aber natürlich variieren Materialien und Mittel von Bau zu Bau, und es gilt, die Regeln den Situationen anzupassen. Jedoch gibt es eine Methode des Denkens, die man in allen Fällen anwenden sollte, und das ist nichts anderes als die Anwendung des praktischen Verstandes.«

Diese Bemerkungen, in denen wir die Stimme Segals zu hören meinen, wurden ein Jahrhundert früher geäußert. Hier spricht »Le Grand Cousin«, der Architekt in E. E. Viollet-Le-Duc's Essay *L'histoire d'une maison*. Viollets faszinierendes Buch ist als Diskurs zu einer Methode zu verstehen, der Methode des gesunden Menschenverstandes, der seine Wurzeln in der Realität des täglichen Lebens hat. Wir werden in diesem Buch mit einem Menschen bekannt gemacht, der mit allen Aspekten des Bauens vertraut ist, mit denen des Materials, der Technik, des Rechts, der Wirtschaftlichkeit und der Ästhetik. Seine Kraft beruht auf konkretem Wissen und dem unmittelbaren Kontakt. Er ist ein Mann des Bauplatzes, dort beherrscht er alle Möglichkeiten der Materialien und Prozesse. Er zeichnet nicht um des Zeichnens willen, sondern nur um Ideen zu erklären, die er vorher durchdacht hat.

Walter Segal hätte in diesem Eugène einen Gleichgesinnten erkannt, hätte die Beschreibung doch auch auf ihn gepaßt.

Viollets Buch von 1875, angeblich für junge Leute geschrieben, wurde als »das beste Buch, das man einem jungen Mann, der an eine Karriere als Architekt denkt, in die Hand geben könne«, bezeichnet (Gazette des Beaux Arts, 1880). Ein Jahrhundert später nannte es Jacques Aillagon ein »ergreifendes Plädoyer für den Beruf des Architekten« (Architectural Design 1980).

Ebenso könnte man Segals Karriere als ein Plädoyer für die Möglichkeiten eines Architekten unserer Zeit bezeichnen.

Early Life

Ascona

Walter Segal's early life is fascinating for the extraordinary context within which he grew. It is equally interesting for the light his »innocent eye« casts on those remarkable melting pots – Monte Verità and Swiss Dada during the 1914–18 war; Berlin in the early twenties; and European architectural ideas in the later twenties. It was formative, and remained vivid for him; he talked about those years at length and with a pertinence which can only be briefly hinted here.

Segal's childhood memories were of life on the edge of Lago Maggiore, at Ascona where Switzerland is almost Italy, on the edge between Monte Verità (a utopian colony set up in 1900) and a small traditional fishing community. Half an hour's walk down steep steps from Monte Verità, the Segal family house became a centre of conviviality, a natural meeting place for all sorts of interesting people whom Monte Verità had attracted.

Friends who visited included Lou-Lou Albert-Lazard from Paris, Hans Arp who regularly came to stay from Zurich (as later did Tristram Tzara and Hans Richter, Hugo Ball and Emmy Hennings, the founders of Dada), and the writers Leonhard Franck and Eli Ludwig. Friends among the Ascona settlers included Henri Oedenkoven (co-founder of Monte Verità); Viking Eggeling (who had been a pupil of Derain and was later to make the first abstract films); Heinrich Goesch and his brother Paul, (»Tancred« in Taut's »Glass Chain« correspondence); »the most extraordinary« Johannes Nohl; and Otto and Adya van Rees from Holland. In 1917, with his three women and his son, there came the painter Alexi Jawlensky who for a time eloped with Segal's mother.

For Walter Segal, the boy of seven to twelve years, it was »a completely unusual childhood, and probably the best sort of gift I had in my life. I, of course, never asked what anything was. You took for granted what you could not comprehend.«

The family friend he spoke of with most affection was Hans Arp, who produced sound poems which he performed for the Segal children's amusement. »My sister and I composed one which Arp promised to publish in ›DADA‹,« recalled the boy. »But he never did, because children, of course, come into a different category and when *they* do this sort of thing it is their habit, whereas with grown-ups it's some sort of insight. Arp's sculptures were the first non-representational works I knew, but in this childlike world I thought I could do better....«

From this hothouse, the boy escaped to village, lake and mountain – »and what an escape! When I think what I might have become!« He always talked of Monte Verità as an outsider – proud to have been a penniless, shoeless, village boy playing truant. (Although natural-

Woodcut of Ascona by Arthur Segal.
Holzschnitt von Arthur Segal: Ascona.

Die frühen Jahre

Ascona

Eine faszinierende und ungewöhnliche Umgebung prägte Walter Segals frühe Jahre: Die Bewegungen des Monte Verità und des Schweizer Dada während des Krieges 1914–18, Berlin in den frühen Zwanziger und die europäischen Architekturströmungen der späten Zwanziger Jahre. Alles das hat ihn geformt und ist in ihm lebendig geblieben; er sprach oft ausführlich über diese Jahre und mit einer Anteilnahme, die hier nur angedeutet werden kann.

Zu Segals Kindheitserinnerungen gehörte das Leben am Lago Maggiore, in Ascona, zwischen dem Monte Verità, wo um 1900 eine utopische Kolonie entstanden war, und einem kleinen traditionellen Fischer-

Woodcut by Walter Segal as child in Ascona.
Holzschnitt von Walter Segal als Kind in Ascona.

dorf. Im Dorf war das Haus der Familie Segal ein Zentrum der Geselligkeit, ein Treffpunkt der verschiedensten interessanten Menschen, die der Monte Verità angezogen hatte. Zu den Freunden, die hier verkehrten, gehörten Lou-Lou Albert-Lazard aus Paris, Hans Arp, der regelmäßig aus Zürich herüberkam, ebenso wie später Tristram Tzara und Hans Richter, Hugo Ball und Emmy Hennings, die Gründer des Dada, und die

Schriftsteller Leonhard Franck und Eli Ludwig. Zu den Freunden, die sich in Ascona angesiedelt hatten, gehörten Henri Oedenkoven (Mitbegründer des Monte Verità), Viking Eggeling (der ein Schüler Derains gewesen war und später die ersten abstrakten Filme drehte), Heinrich Goesch und sein Bruder Paul, »Tancred« in Tauts Glas-Ketten-Korrespondenz, der »sehr eigenartige« Johannes Nohl sowie Otto und Adya van Rees aus Holland. 1917 kam, mit seinen drei Frauen und seinem Sohn, der Maler Alexi Jawlensky, der für eine kurze Zeit mit Segals Mutter durchbrannte.

Für Walter Segal, den Jungen zwischen sieben und zwölf Jahren, war es eine »absolut ungewöhnliche Kindheit, und vielleicht die beste Gabe, die das Leben für ihn hatte. Ich fragte natürlich niemals, was das alles war. Man nahm als gegeben, was man nicht begreifen konnte.«

Mit der größten Zuneigung sprach er von Hans Arp, der Klang-Gedichte produzierte, und sie zum Vergnügen der Segal-Kinder vortrug. »Meine Schwester und ich produzierten auch eines, das er versprach im Dada zu veröffentlichen«, erinnert sich der Junge, »aber er hat es niemals getan; denn Kinder gehören natürlich in eine andere Kategorie. Wenn sie solche Sachen produzieren, ist es eben ihre Art, während es bei Erwachsenen einer höheren Einsicht entspringt. Arps Plastiken waren die ersten nicht-gegenständlichen Kunstwerke, die ich sah; aber in dieser meiner kindlichen Welt dachte ich, das könne ich besser ...«

Aus dieser Treibhausathmosphäre flüchtete der Junge ins Dorf, an den See und in die Berge. »Und was für eine Flucht. Wenn ich denke, was aus mir hätte werden können.« Vom Monte Verità sprach er immer als ein Außenseiter, stolz, daß er, der barfüßige Dorfjunge, ohne einen Pfennig, die Schule geschwänzt hatte. Obwohl er von Natur aus begabt war, haßte er es zu lernen, und die Idee, routinemäßig zu arbeiten, nach der Uhr, hat er niemals akzeptieren können.

Seine Erziehung fand im Freien statt. Er baute mit Holz und Steinen und lernte, Pläne zu zeichnen. »So glitt ich langsam in ein Verständnis für das Bauen, und

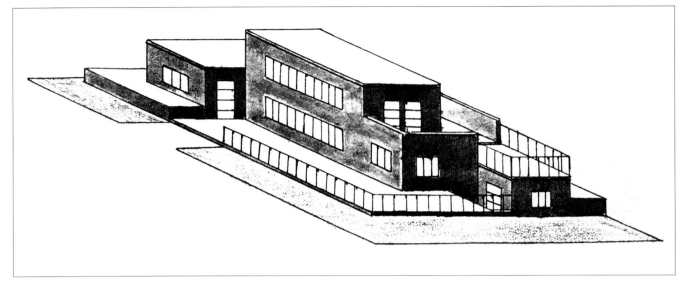

Segal's first schoolboy design: private house 1924/25.
Segals erster Entwurf: Als Schuljunge zeichnete er ein Einfamilienhaus, 1924/25.

ly gifted, he hated study and the idea of working to routine, disciplined by a clock, is one he never ever came to accept.)

His happiest education was out of doors; he constructed in wood and stone and learned to draw plans. »So I gradually slid into an understanding of how buildings are put up, and it was clear to me by the age of fourteen that I was going to be an architect. This combination of designing and making was extremely important for me.« Oedenkoven's buildings on Monte Verità were clearly imprinted on the boy – flat roofs with wide overhangs, roof terraces, large sliding windows and doors, and a use of natural and untreated materials.[1]

»This then was the world of my childhood. I took it for granted and questioned it with what feeble powers I possessed. A childhood in Ascona provided a background on to which I could fall back in years to come.«

Berlin

But the war ended and his parents' friends dispersed. The van Reeses to Paris, Arp to Meudon, and Tzara was having Loos build him a house in Paris. Others had gone to Berlin with high hopes in the wake of the social-democratic revolution. Thither headed the Segals, in 1920 finally returning to their old flat in Charlottenburg 5, Berlin, just after Walter's thirteenth birthday. After

those years in Ticinio the youth had to be dragged, reluctantly, to Berlin.

His father joined the November Group (which had been formed as a consequence of the 1918 Arbeitsrat für Kunst), and before the end of the year was on the committee and at the centre of the Berlin artistic milieu. Mies van der Rohe, Häring and Hilberseimer were members, and indeed most of the post-war avant-garde artists – from Schwitters and Dadists to the Bauhaus-meisters – belonged for some time at least.

The Segal »open house« and their monthly *jour fixe* became famous in the city as avant-garde meetings. »There was Gropius and Behne, Moholy and Kemeny, Mies who never said a word and Hilberseimer who spoke a lot, and, of course, painters sculptors and writers galore, including such characters as Kurt Schwitters and Raoul Hausmann. One day there sat in a corner Kasimir Malevich, the founder of the legendary suprematists, who bitterly complained to all willing to listen how Moholy and Lissitzky had stolen from him....« All their tales, and especially the stories exchanged at such meetings about Bauhaus exercises, diet and life under Itten led the youth to comment: »Purest Ascona, Germanic version, and consequently deadly bitter and without a trace of humour.«

It was a remarkable milieu to grow up in. Money

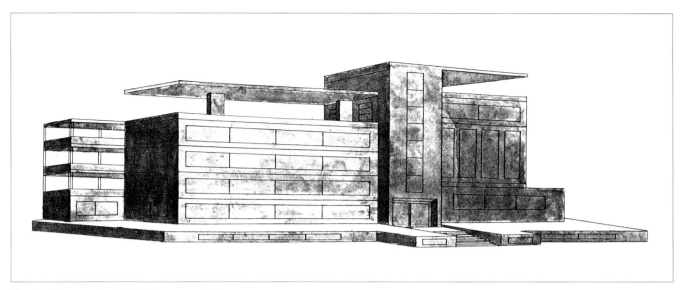

Design for a school 1925, perspective.
Mit 18 Jahren entwarf Segal eine Schule.

im Alter von Vierzehn war es schon klar, daß ich Architekt werden wollte. Die Verbindung von Entwerfen und Ausführen war für mich außerordentlich wichtig.« Oedenkovens Bauten auf dem Monte Verità prägten sich dem Jungen klar ein – Flachdächer mit weitem Überstand, Dachterrassen, große Schiebefenster und Schiebetüren, die Verwendung natürlicher, nicht behandelter Materialien.

»Dieses war die Welt meiner Kindheit. Ich nahm sie als gegeben und stellte sie nur mit meinen damaligen schwachen Kräften in Frage. Eine Kindheit in Ascona gab den Hintergrund, auf den ich mich in kommenden Jahren zurückbesinnen konnte.«

Berlin

Als der Krieg zu Ende war, zerstreuten sich die Freunde der Eltern. Die van Rees zogen nach Paris, Arp ging nach Meudon, Tzara ließ sich von Loos ein Haus in Paris bauen. Andere waren nach Berlin gegangen, mit großen Hoffnungen in die sozialdemokratische Revolution. Dahin zog es auch die Segals. Kurz nach Walters dreizehntem Geburtstag kehrten sie in ihre alte Wohnung in Berlin-Charlottenburg zurück. Nach den Jahren im Tessin ließ sich der Junge nur mit Widerwillen nach Berlin zurückschleppen.

Segals Vater schloß sich der sog. November

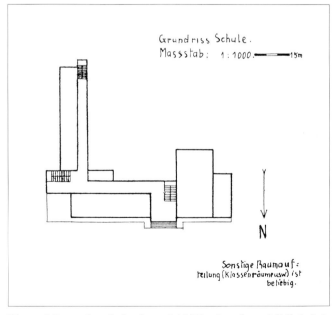

Plan of the school, designed 1925, showing childish lettering.
Plan der 1925 entworfenen Schule, mit Originalbeschriftung.

was scarce and inflation raging; his father desperately tried to sell paintings and teach pupils who could afford to pay, but with little success. By August 1923, one million German marks bought one US dollar.

The youth understood clearly: »I discovered when I was fourteen, that there was absolutely no money. I knew then that I would have to get a scholarship; and I *hated* school!«

Walter Segal, in fact a brilliant student, acquired in these years a foundation for his lifelong love of history and of language. Throughout his life he remained meticulously discreet about his learning; very occasionally would he quote chunks of Homer in Greek and only slightly more often from the Latin classics.

The king's new clothes

From a remarkably young age, Walter Segal showed a mature ability not to be bowled over by the fashionable notions or popular creeds. He was never prepared to accept a consensus value which society might give anyone. »I was one of the children that were produced by the Avant Garde in the arts at the beginning of the Century. If you did not find a way to escape from this milieu, it would engulf you. And that was what I instinctively sensed as a child.« Segal's passionate scepticism therefore began early – as self-preservation.

»I never committed myself [to factions]. It was interesting, and isolating, but the only way that I could exist. This means that I was not accepted by the right or by the left as one of theirs. I think my sanity depended on that.... I have remained an observer, to this day. I am intensely curious in observing how people live. But as an observer, I had to face non-acceptance in all the different camps. I have always been a fierce individualist.« He developed an acute nose for pretention, for bluff, and for jargon. The mature Walter Segal could sniff not just an Achilles' heel but, more usefully, could sense a clay foot from a hundred paces. The childlike common-sense, the »seeing through« to the reality, began early: where others saw the utopian vegetarianism of Monte Verità, the boy Segal looked at those living on fruit and raw vegetables and saw all kinds of digestive illnesses.

For today's distant generation, this undivertable, clear vision also makes Segal an exemplary recorder of that period; in his personal experience, producing snapshots in sharp, cool focus. By the 1950s, Segal had a quiet reputation as »a mocker of the grey eminences of the modern movement« (in Reyner Banham's words[2]). Though Segal coyly denied it – »I am but one of their seemingly renegade offsprings« – he was much more than that. He was a clarifying lens on whatever his glance had fallen.

Segal felt that as a child he had seen the lack of substance in the rhetoric around him. In fact, with that particularly acute adolescent sensitivity to inauthenticity, he saw the lack of coincidence between the public man and the private person; the test of privately practicing what is publicly preached. But, further, there are the claims that architecture can have moral values: »honest detailing« and »sincere building« are phrases so ingrained over two centuries that we hardly notice them. The contrast between such claims and the flawed lives of the claimants can be painfully obvious; and here it was the moralising to which Segal objected. »A passionate moralist like Le Corbusier could brand a minor design decision a sin,« he said. The other contrast in his gaze, and one obvious in so much of the twentieth century avant-garde, is the distance between obfuscating rhetoric and the works produced.

»These people were incessantly talking about ART, even at mealtimes – very boring indeed.« But the adult Walter Segal still saw the word art written as ART, slightly to mock its pretention, but also a bit timorous of its power, shying from questions of the meaning of the work itself. The clarity of the child sees the foibles but not what the artists are trying to express in feelings and ideas within their work. Does it knock on doors too private for its emotional intrusion? »Trusting oneself to artistic impulses is dangerous and unhingeing,« Segal warned. But if the adult Segal only plays the child, mute rather than dumb, the potential child in all of us is touched by how often, accurately and humiliatingly he manages to point out the naked king.

Others have seen Segal's enviable ability to retain a childlike perception as his own clay feet in evidence. Certainly it allowed him throughout his life, when he wished, to ignore inconvenient contradictions; it encouraged the prankster and also the charmer; and it invited mothering from those who would care for him.

Go-betweens

While he »saw through« by instinct, there remained those few whom Segal did admire: certainly Bruno Taut, perhaps Oud, Stam, maybe Mendelsohn and a few others; and of course his own father. All people whose

Gruppe an, und nach wenigen Monaten war er Mitglied des Komitees und stand er im Mittelpunkt des Berliner künstlerischen Milieus. Mies van der Rohe, Häring und Hilberseimer gehörten zur Gruppe wie die meisten Avantgarde-Künstler der Nachkriegsjahre – von Schwitters und den Dadaisten bis zu den Bauhaus-Meistern, zumindest für eine gewisse Zeit.

Arthur Segals offenes Haus und der monatliche »Jour fixe« wurden als Avantgarde-Treffpunkte in der Stadt berühmt. »Da waren Gropius und Behne, Moholy und Kemeny, Mies, der niemals ein Wort sagte, und Hilberseimer, der um so mehr redete, und natürlich Maler, Bildhauer und Schriftsteller in Hülle und Fülle, einschließlich Charaktere wie Kurt Schwitters und Raoul Hausmann. Eines Tages saß in einer Ecke Kasimir Malevich, der Begründer der legendären Suprematisten, der sich bitterlich beklagte bei jedem, der es hören wollte, wie Moholy und Lissitzky von ihm gestohlen hätten ...«

Alle ihre Erzählungen, besonders die Geschichten über Bauhaus-Übungen, Diät und Leben unter Johannes Itten, entlockten dem Jüngling den Kommentar: »Reinstes Ascona, germanische Version, todernst, ohne eine Spur von Humor.«

Das Geld war knapp, die Inflation erreichte Höhepunkte. Der Vater versuchte verzweifelt, Bilder zu verkaufen und Schüler zu unterrichten, die es sich leisten konnten zu bezahlen, aber mit wenig Erfolg. Im August 1923 kostete ein US-Dollar eine Million Mark.

Der Junge sah das ganz klar: »Ich entdeckte, als ich vierzehn war, daß wir absolut kein Geld hatten und begriff, daß ich ein Stipendium bekommen mußte, und ich haßte die Schule.«

Des Kaisers neue Kleider

Schon in jungen Jahren zeigte Segal die Fähigkeit, sich nicht durch modische Anschauungen oder populäre Glaubensbekenntnisse überwältigen zu lassen. Er war niemals gewillt, sogenannt allgemein anerkannte Werte unbesehen zu akzeptieren. »Ich war eines dieser Kinder, die die künstlerische Avantgarde zu Beginn des Jahrhunderts produziert hatte; wenn man keinen Weg fand, diesem Milieu zu entfliehen, würde es einen verschlingen. Und das habe ich als Kind intensiv gespürt.« Segals leidenschaftliche Skepsis hat frühe Wurzeln. Sie war ein Stück Selbsterhaltung.

»Ich verpflichtete mich niemals (einer Partei). Das isolierte, aber war die einzige Art für mich, zu existieren. Es hieß, daß ich weder von der Rechten noch von der Linken als einer der ihren akzeptiert wurde. Ich glaube, meine geistige Gesundheit hing davon ab ... Bis heute bin ich ein Beobachter geblieben. Ich bin ungeheuer neugierig zu beobachten, will wissen, wie Leute leben. Aber als Beobachter hatte ich mich damit abzufinden, keinem Lager anzugehören. Ich bin immer ein wilder Individualist gewesen.«

Er entwickelte eine feine Witterung für Anmaßung, Bluff und Redensarten. Der reife Walter Segal würde nicht nur eine Achilles-Ferse entdecken, sondern, wichtiger noch, einen Klumpfuß aus hundert Schritt Entfernung erkennen. Sein kindlicher »gesunder Menschenverstand«, seine Art, die Dinge zu durchschauen, bildeten sich sehr früh aus: Wo andere das utopische Vegetariertum am Monte Verità verehrten, stellte der kleine Segal bei allen, die nur von Obst und rohem Gemüse lebten, Verdauungsbeschwerden fest.

Für uns wird Segal darum zu einem einmaligen Chronisten jener Zeit. Aus eigener Erfahrung vermittelt er so etwas wie scharfe, das Objekt kühl erfassende Momentaufnahmen. Während der 50er Jahre hatte Segal den Ruf, er sei »ein Spötter vor den Grauen Eminenzen der Modernen Bewegung« (Reyner Banham, The Architects' Journal 7. 6. 1956, S. 630). Segal leugnete dies, wie es sich gehört: »Ich bin nur einer ihrer anscheinend abtrünnigen Nachkommen.«

Segal erkannte schon als Kind die mangelnde Substanz in all der Rhetorik rings um ihn herum. Der Heranwachsende war empfindlich auf Mängel an Übereinstimmung zwischen Rollen- und Lebensverhalten. Die Forderung, privat so zu leben, wie man es öffentlich predigt, war für ihn selbstverständlich. Darum bemerkte er sofort den Kontrast zwischen dem Anspruch, daß Architektur einen moralischen Wert haben solle, und dem Leben derer, die ihn stellten. Segal irritierte es, daß »ein leidenschaftlicher Moralist wie Le Corbusier eine unwesentliche Entscheidung in der Gestaltung als Sünde brandmarken konnte«. Ebenfalls sah er bei vielen der Avantgarde des 20. Jahrhunderts, daß ihre verwirrende Rhetorik in keinem Verhältnis zu ihrer tatsächlichen Leistung stand.

»Diese Leute sprachen unaufhörlich über KUNST, selbst beim Essen – wirklich sehr langweilig.« Er machte sich ein wenig lustig über den Anspruch, andererseits schreckte er ein wenig zurück vor den Fragen nach der Bedeutung des Werkes. Das Kind sah mit klarem Blick die schwachen Seiten, nicht aber, was der Künstler an

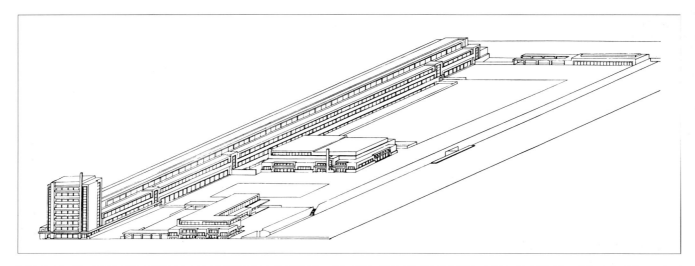

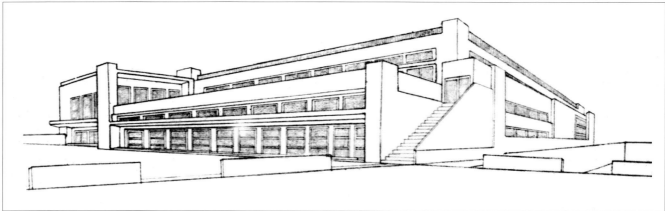

work changed, whose reputations fluctuated widely, and who – perhaps quite unfairly – have become footnotes in the smooth text of those times.

Of a 1928 Stam building Segal wrote: »The immaculate detail is still visible today. Beyond this there is a new feeling for space, an ease and lightness which invited comparison; its elegance was natural and quite free from over-emphasis on visual accents already clouding the minds of other designers.« Segal comes close to a definition of his *own* goals in building.

Even were we to lengthen his list, apart from the soon-to-be-exiled Mendelsohn, there would remain only one German. To Segal, Germans were »a heavy-minded race«, and he was appalled by the Romantic legacy which he saw spilling with an awful logic from the early 19th century towards Nazism and beyond.

Competition design for a large exhibition hall, entered by Segal in his last year at school, 1926.

Wettbewerbsentwurf für eine große Ausstellungshalle, gezeichnet von Segal in seinem letzten Schuljahr, 1926.

Gefühlen und Ideen auszudrücken sucht. »Sich selbst künstlerischen Impulsen hinzugeben, ist gefährlich und beunruhigend«, warnte Segal sein Leben lang.

Der erwachsene Segal liebte es, das Kind zu spielen, womit er das potentielle Kind in uns allen anspricht. Er zeigt immer wieder augenzwinkernd auf die Nacktheit des Kaisers.

Wir, seine Freunde, haben Segals beneidenswerte Fähigkeit erkannt, kindliche Wahrnehmungen beizubehalten und zu zeigen. Das erlaubte ihm sein ganzes Leben lang, unbequemen Widerspruch zu ignorieren; es ermutigte den Zauberer und den Charmeur in ihm; und es weckte beschützerische Instinkte bei denen, die ihn mochten.

Einen Maßstab anlegen

Während er instinktiv »durchschaute«, bl eben doch einige wenige, die Segal bewunderte: gewiß Bruno Taut, vielleicht Oud, Stam, Mendelsohn und ein paar andere, und natürlich sein eigener Vater. Alle sind Leute, deren Werk Schwankungen und deren Ruf einem Auf und Ab unterworfen war, und die – vielleicht ganz zu Unrecht – in den gängigen Büchern über diese Zeit zu Fußnoten geworden sind. Über einen Bau von Stam aus dem Jahre 1928 schrieb Segal: »Die fehlerfreie Detaillierung ist noch heute sichtbar. Darüber hinaus gibt es ein neues Gefühl des Raumes, eine Leichtigkeit und Helle, die zu Vergleichen herausfordert; die Eleganz war natürlich, frei von der Überbetonung visueller Akzente, die schon damals den Geist mancher Entwerfer umwölkte.« Das kam nahe an eine Definition seiner eigenen Ziele.

Unter den von ihm Hochgeschätzten gab es, abgesehen von Mendelsohn, der schon bald ins Exil ging, nur einen Deutschen. Für Segal waren Deutsche eine »schwerblütige Rasse«. Und das Schwärmerische, das sich in einer fürchterlichen Logik vom frühen 19. Jahrhundert bis in den Nazismus und darüber hinaus fortentwickelte, schreckte ihn. Die Ausnahme war Bruno Taut. Viel später schrieb Segal über ihn:

»Gerade weil er neben der Hauptströmung der ›Moderne‹ jener Jahre daherschritt, nicht in ihrer Mitte, weil er kollektive Aktionen predigte, sich jedoch nie in ihnen verlor, gewann er am Ende mehr, als er damals zu verlieren schien.« »Ich stelle fest, daß Bruno in keine Gemeinschaft hineinpaßt«, schrieb 1919 sein enger Freund Gropius.

»So unkonventionell wie in seiner Arbeit, war er

auch im Leben. Im persönlichen Kontakt verfolgte er keine Mission, auch war er nicht so erschreckend respektheischend. Das erhöhte das Vergnügen an seiner Gesellschaft; im Kontakt mit jungen Leuten war er ganz natürlich, ungezwungen, keine Spur von einem Hohepriester, einem Führer, einem Lehrer, kein Bedarf an Weihrauch, an intelligenten Fragen; Taut liebte Gleichberechtigung ... Er hatte einen starken Sinn für Humor und ein grenzenloses Interesse für Menschen. Um ihn herrschte eine offene Atmosphäre. Seine Respektlosigkeit war ein Labsal. In seinem Haus traf man andere aus einer anderen Generation, die sich jedoch nicht in Container eingeschlossen hatten.«

Es lohnt sich, diesen Tribut ausführlich zu zitieren, zeigt er doch, nach welchen Qualitäten Segal selbst in seinem Leben und seiner Architektur strebte.

Der andere zentrale Einfluß ging von seinem Vater, Arthur Segal, aus.

Die Familie

Seine Eltern Arthur Segal (geb. 1875) und Ernestine Chavas (geb. 1879) waren rumänische Juden, Vetter und Cousine. Ernestine kam mit 17 Jahren mit ihren Eltern nach Berlin. Arthur lief mit 17 seiner wohlhabenden bürgerlichen Familie davon, zunächst nach München, dann nach Berlin. Seine sozialistischen Neigungen und seine künstlerischen Absichten verlangten nach einem weiteren Horizont, als ihn die kleine Provinzstadt in Rumänien bieten konnte. (Siehe den Katalog Arthur Segal, 1875–1944 von Wolf Herzogenrath und Paven Liska, Argon-Verlag, Berlin, 1987.) Ernestine und Arthur heirateten 1904 in Berlin. Walter Segal wurde 1907 geboren, seine Schwester ein Jahr später.

Arthur Segal entwickelte eine Lehre der »Gleich-Wertigkeit«, die für ihn sowohl für die Komposition seiner Bilder als auch für seine sozialistischen politischen Ideale von Bedeutung war. »In meiner Kunst suche ich einen Weg, ausgewogene Werte auszudrücken, das heißt Dinge, die von gleichem Wert sind ...« Er fing an, Malerei zu unterrichten, eine Begabung, die er durch sein ganzes Leben weiterentwickelte, und die ihn in vielen Aspekten als einen Pionier der künstlerischen Therapie zeigt, obwohl er wohl niemals diesen Ausdruck gebraucht hat. Freud bestätigte das am 20. September 1937 von Wien aus, als Arthur Segal ihn um eine Referenz anläßlich der Eröffnung seiner Malschule in England bat.

That exceptional German among Segal's influences was the most important: Bruno Taut. Much later Segal wrote:

»In marching alongside the mainstream of the ›modern‹ movement in those years but not within it, Taut preached collective action but was never submerged by it, and in the end gained more than he seemed to lose at the time. ›I perceive Bruno does not fit into any community‹, wrote his once close friend Gropius in 1919.

»Unconventional in his work,« Segal continued, »[Taut] was not hidebound in life either. In personal contact he did not embody a mission, nor was he formidable and demanding of respect. That increased the pleasure of being in his company. His contact with young people was natural; with them he was completely at ease. No touch of the high-priest, the leader, the teacher. No need for frankincense, for ›intelligent‹ questions; Taut loved equality....

»He had a considerable sense of humour and boundless interest in people. There was an open atmosphere about him. His irreverence was a tonic. At his house one met others, not of one's own generation, who had not put themselves into containers.«

This tribute is worth quoting at length as indeed it could describe the qualities Walter Segal aspired to in his own life and architectural practice.

The family

The one other central influence was his father Arthur.[3] His parents, Romanian Jewish cousins, met in Berlin and married in 1904, in which year Arthur Segal exhibited paintings at the New Secession there. Walter Kurt Segal was born in 1907 and his sister a year later.

Arthur Segal developed a theory of »equal- value« which had significance for him from the layout of a painting to his socialist political ideals. »My art wants to find a way through expression of ›equivalent value‹, of things being equally valuable....« He began to teach painting, a talent he developed all his life, and which in many respects shows him as a pioneer of art therapy.[4]

Themes of rootlessness and financial insecurity recur through their lives, as they develop an ability to slip unnoticed into the interstices between great events – a skill Walter Segal never lost. They left Berlin in 1914 for Ascona where, in 1916, their status turned from tourists to political refugees. After the war, the Segals had been back in Berlin 13 years when Hitler became

Chancellor, and they quickly moved again. They joined Walter in Palma de Mallorca, staying until the Fascists gained power in Spain, and then finally escaping to England.

Father and son

Walter Segal described his father as a warm, lovable innocent. »But I had no filial piety for either the man or his paintings.« In fact he had little time for his father's intellectually limited philosophising, and not much more for »abstract« art. He must really have felt the weight of the paradox that unfolded as his father's work from the 1930s matured towards a genuinely personal, figurative vision – and his reputation simultaneously vanished into eclipse.

In the 1930s the father wrote of his problem in developing an understanding with his son Walter, by then an independent architect. »Our views are extremely different,« he wrote. Once in Mallorca, he was relieved when Walter left for a year in Egypt. »It's a good thing he is a long way away. I simply hope never to be forced to rely on him. And I hope he will never have to rely on me.«

Very few years later, Walter was hard at work organising for the arrival of his family in Britain, and then for the foundation of his father's new painting school in London.

Becoming an architect

Even as a schoolboy, Walter Segal had no difficulty in mimicking images of the new architecture. He had produced designs which were Dudokish (after a youthful visit to Holland with his father), and cardboard-like reminiscences of his father's Berlin friends (among whom, he said, none was more ludicrously easy to mimic than Ludwig Hilberseimer).

»At about this time Behne told my mother that I seemed to have found an easy access to the forms of modern architecture as it appeared then to be, without really understanding it. And, of course, he was right for I hardly could make plans and I was tortured by not having a clue of how such things could be built.« Nothing daunted, the schoolboy next entered an architectural competition with a design for a vast exhibition hall with office block. »For my finals at school I had to compose a thesis. It had the high-sounding title, *In what way does modern architecture answer the aesthetic concepts of our time.* Not that I knew anything about

Wurzellosigkeit und finanzielle Unsicherheit begleiteten die Segals durch ihr ganzes Leben. Sie gingen 1914 von Berlin nach Ascona, wo sich ihr Status 1916 von Touristen zu politischen Flüchtlingen wandelte. Nachdem sie dann nach Berlin zurückgekehrt waren, dauerte es nur 13 Jahre, bis Hitler Kanzler wurde. Hierauf zogen sie schnell weiter, und zwar zu Walter nach Palma de Mallorca, wo sie blieben, bis die Faschisten in Spanien die Macht erlangten. Von dort entkamen sie nach England.

Vater und Sohn

Walter Segal beschreibt seinen Vater als einen warmherzigen, liebenswerten und unschuldigen Menschen. »Aber ich hatte kein kindliches Mitgefühl, weder mit dem Mann noch mit dem Maler.« Er hatte auch wenig Geduld für die intellektuell begrenzte Philosophie seines Vaters und nicht viel mehr für die abstrakte Kunst. Er muß als paradox erkannt haben, daß das Werk seines Vaters nach 1930 zu einer ganz persönlichen Kunst figurativer Visionen reifte, womit aber sein Ruf verklang.

1930 schrieb der Vater über sein Verhältnis zu seinem Sohn Walter, der damals schon ein selbständiger Architekt war: »Unsere Ansichten sind diametral entgegengesetzt.« Während der Zeit in Mallorca war er erleichtert, als Walter für ein Jahr nach Ägypten ging. »Es ist gut, daß er so weit weg ist. Ich hoffe nur, daß ich niemals gezwungen sein werde, mich auf ihn zu stützen, oder er sich auf mich.«

Wenige Jahre später organisierte Walter den Empfang der Familie in England und dann die Eröffnung der neuen Malschule seines Vaters in London.

Segal wird Architekt

Schon als Schuljunge hatte Segal keine Schwierigkeiten, die großartigen Darstellungen der neuen Architektur nachzuahmen. Er fertigte Zeichnungen nach der Art der Berliner Freunde seines Vaters an. »Ungefähr um diese Zeit erzählte Behne meiner Mutter, daß ich anscheinend einen leichten Zugang zu den Formen der modernen Architektur gefunden habe, so wie sie sich darstellen, ohne sie wirklich zu verstehen. Natürlich hatte er recht; denn ich konnte ja kaum einen Plan zeichnen, und es quälte mich, daß ich keine Ahnung hatte, wie man so etwas baut.« Ganz unerschrocken beteiligte sich jedoch der Schuljunge an einem Archi-

tekturwettbewerb, den Entwurf für eine große Ausstellungshalle mit Bürogebäude.

»Für meine Abschlußprüfung in der Schule hatte ich einen Aufsatz zu schreiben. Er trug den hochtrabenden Titel ›In welcher Weise entspricht die moderne Architektur dem ästhetischen Konzept unserer Zeit?‹ Nicht, daß ich irgendetwas über moderne Architektur gewußt hätte oder auch über Ästhetik; aber ich legte meine Ansichten vor, und sie sind heute noch sehr lustig zu lesen. Den Rest suchte ich zusammen, in einer Taktik des offenen Diebstahls aus Behnes *Der moderne Zweckbau*, von dem ich ein Exemplar zusammen mit dem Manuskript einreichte.«

Er erhielt die besten Noten und ausgezeichnete Empfehlungsschreiben von Freunden der Familie, wie Gropius und Erich Mendelsohn, und bekam dann auch ein Stipendium für ein Architekturstudium irgendwo in Europa mit Studiengebühren und Lebenskosten. »So mußte ich nun entscheiden, wohin ich gehen wollte.«

»Es war ein glücklicher Umstand, daß meine Mutter und mein Vater viele Architekten kannten, die damals der modernen Architektur zum Durchbruch verhalfen. Mein Vater war ein enger Freund von Oud und kannte auch Rietveldt, war befreundet mit Berlage. Anderseits waren da Kontakte zu Gropius und allen Malern am Bauhaus.«

Als Gropius die Segals im Winter 1926 besuchte, ermutigte er Walter, ans Bauhaus zu kommen. Dieser reagierte mit folgenden Fragen: »Erstens, gibt es da einen Kurs über die Theorie der Konstruktionen? Und als er antwortete, nein, haben wir nicht, aber wir hoffen, einen Ingenieur anstellen zu können, war das schon ein Minuspunkt. Zweitens, unterrichten sie Geschichte? Nein, sagte Gropius. Er glaube nicht, daß es für einen modernen Architekten notwendig sei, etwas von Geschichte zu wissen. Aber ich sagte, ich wolle wissen, wie die gothischen Kathedralen konstruiert seien. Das war der zweite negative Punkt. Und dann erzählte mir Andreas Feininger auch noch, daß ich am Bauhaus gar nichts lernen würde. Zum Schluß entschloß ich mich für die übliche Ausbildung, obwohl mir davor graute ...«

Zuerst ging er nach Holland; die Schule dort war reaktionär; aber es gab lebhafte Debatten und viele neue Bauten. Durch Oud, den Freund der Familie, lernte er die *De Stijl* Gruppe kennen, und in Amsterdam gab es den Romantizismus, den Segals Freunde »zum

modern architecture, or aesthetics for that matter, but I propounded my views which are still a scream to read; the rest I culled, by using the tactics of daylight robbery, from Behne's *Der Moderne Zweckbau* a copy of which I submitted with my script.«

He achieved top grades, got excellent letters of recommendation from family friends Walter Gropius and Erich Mendelsohn, and was granted a scholarship to pay all tuition and living costs while he studied architecture anywhere in Europe. So he had to decide where to go.

»It was fortunate that my mother, and my father, knew a large number of architects then introducing modern architecture. My father was a close friend of Oud, he also knew Rietveldt, he was friendly with Berlage. On the other hand, there were contacts with Gropius and, naturally, with all the painters of the Bauhaus.« Gropius, while visiting the Segals in the winter of 1926, tried to encourage Walter to join the Bauhaus. He responded with these questions: »First, had they a course on theory of structures? And he said: no, they hadn't; but they hoped to engage an engineer. So that was a black mark against him.

»Second, were they teaching history? No, Gropius said. He didn't think that for a modern architect it was necessary to know anything about history. But I said I wanted to know how the gothic cathedrals were constructed. So that was the second black mark. That, on top of Andreas Feininger telling me that I wouldn't learn anything at the Bauhaus! In the end I made up my mind to get an ordinay education, though I dreaded it.«

First he went to Holland where, although the Delft School under Grandpré Moliere was reactionary, there was lively debate and much new building. He met the De Stijl group through his family friend Oud, and in Amsterdam saw the »romanticism« which Segal's colleagues found »screechingly funny.« Isolated in Hilversum was Dudok, whose buildings Segal had looked at with more interest, and then there were the even less fashionable Bijvoet, Stam and Duiker whom Segal came particulary to admire. It was all very confusing.

He soon decided to move on, as his scholarship permitted; and having visited Corbusier's and Lurçat's offices in Paris, he returned to Berlin. At least there a new architecture, if not as outspoken as the Dutch, was beginning. Poelzig and most fluently Mendelsohn were building. And, just when Segal had his fill of masters, he met Bruno Taut.

Student in Berlin

Segal enrolled at the Technical High School in Charlottenburg-Berlin, where he soon upset the traditionalist calm[5]. He was bored by school projects but, entering competitions instead, he began to enjoy himself. He won various prizes which were useful financially but particularly meant »you couldn't really be talked down to.«[6]

The professor who allowed this radical self-education was Hans Poelzig. »I managed to slink into Poelzig's seminar,« Segal said (for it was the most prestigious); and here he thrived. Poelzig's teaching methods, the most progressive in Europe at that moment, were talked about at length by Segal and have been well described[7]. It was, as Segal said, »quite an excellent experience.«

As one of Segal's colleagues at the time explained: »First, he wanted us to do ›our own thing‹, that was Poelzig's phrase; and second, he wanted students to study several alternative solutions. Of course the first was the only possible course for Walter, who always knew what he wanted; while the second, as a design method, appealed to Walter who did it in an *exhaustive* manner!«[8]

Already Segal's working method was astonishingly clear. »Walter would look into a project,« explained a contemporary, »and decide that there were only ten configurations possible. He would work on each, then eliminate seven and develop two in addition to the one he'd chosen as his preferred solution. Then he might hold his own private review with students, before the session with Poelzig, to check all the arguments for and against each scheme. This was very labour-intensive, of course, but Walter had to go at it, had to prove that the way he wants to do it is the only one. Utterly systematic!

»Even then, he had a circle of younger admirers who would assist him with the inking in of drawings, or when a systematic assessment was required. This was essential for his very labour-intensive way of working! He was a perfectionist.«[9]

»He never let himself off lightly,« adds another colleague; »he impressed me even then with his self-discipline. Perhaps he inherited or learned this toughness from his father.«[10]

The systematic attempt to exhaust all design possibilities, which marked a life-long approach, was taken to an extreme when, as a student, Segal submitted with

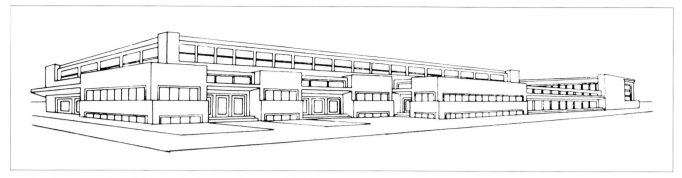

Design for a small exhibition hall with cinema, 1926.
Entwurf für eine kleine Ausstellungshalle mit Kino, 1926.

Schreien komisch« fanden. In Hilversum arbeitete ganz isoliert Dudok, dessen Bauten Segal mit mehr Interesse betrachtete. Und dann waren da noch, weniger im Trend, Bijvoet, Duiker und Stam, den Segal besonders bewunderte. Es war alles sehr verwirrend. Er entschloß sich bald, weiterzuziehen, was ihm sein Stipendium erlaubte. Nachdem er Le Corbusiers und Lurcats Büros besucht hatte, kehrte er nach Berlin zurück. Dort gab es doch den Beginn einer neuen Architektur, wenn auch nicht so offen wie in Holland. In Berlin bauten Poelzig und Mendelsohn. Und, als er genug hatte von den Meistern, traf er Bruno Taut.

Student in Berlin
Segal schrieb sich an der Technischen Hochschule in Berlin-Charlottenburg ein und störte dort bald die traditionelle Ruhe mit einem sehr aggressiven Artikel über die Zustände an dieser Hochschule; er hatte den Text mit zwei Kollegen – einer davon war Julius Posener – zusammengestellt, und der damalige Redakteur der Bauwelt, Friedrich Paulsen, veröffentlichte ihn mit Freude. Die Aufgaben an der Hochschule langweilten Segal, und er fing an, sich an Wettbewerben zu beteiligen; das machte ihm Spaß. Er bekam sogar einige Preise, die finanziell nützlich waren, aber in erster Linie stärkten sie sein Selbstwertgefühl (3. Preis in einem Studentenwettbewerb 1929; ein zweiter Preis in einem Studentenwettbewerb 1930 wurde in der Bauwelt 29/1930/S. 992, veröffentlicht).

Der Professor, der diese radikale Selbst-Ausbildung gestattete, war Hans Poelzig. »Ich brachte es fertig, mich in Poelzigs Seminar zu schleichen« (es war das angesehenste), und hier konnte er gedeihen. Über Poelzigs Unterrichtsmethoden, damals die progressiv-

sten in Europa, hat Segal oft gesprochen, und sie sind gut beschrieben worden (Julius Posener in The Architectural Review 6/1963, S. 401). Es war, wie Segal sagte, »eine ganz ausgezeichnete Erfahrung«.

Einer von Segals Kollegen aus jener Zeit erzählte: »Zuerst sollten wir ›unsere eigene Sache machen‹, so etwa drückte sich Poelzig aus, und, zweitens, sollten die Studenten verschiedene alternative Lösungen studieren. Das erste war natürlich der einzige mögliche Weg für Walter, der immer wußte, was er wollte, während das zweite, als Entwurfsmethode, ihm sehr gefiel, und er nutzte sie extensiv.« (Otto Koenigsberger im Gespräch mit dem Autor, März 1988)

Segals Arbeitsmethode war damals schon erstaunlich klar ausgebildet. »Walter studierte die Aufgabe genau und entschied dann, daß es nur zehn mögliche Lösungen gebe. Diese bearbeite er alle, sieben schied er nachher aus. Zu der einen, der er den Vorzug gab, bearbeitete er noch zwei andere weiter. Dann hielt er seine eigene private Besichtigung mit anderen Studenten ab, um schon vor der Kritik mit Poelzig alle Argumente für und gegen seine Lösung zu prüfen. Das war natürlich sehr arbeitsintensiv, aber Walter mußte sich daran machen, hatte zu beweisen, daß der von ihm gewählte Weg der einzige sei. Durch und durch systematisch. Schon damals hatte er einen Kreis junger Bewunderer, die ihm beim Ausziehen der Zeichnungen mit Tusche halfen, oder bei einer Kostenschätzung, falls das nötig war. Das war die Voraussetzung für dieses arbeitsintensive Entwerfen. Er war ein Perfektionist ...« (ibid)

»Er machte es sich niemals leicht«, fügt ein anderer Kollege hinzu, »er hat mich schon damals mit seiner Selbstdisziplin beeindruckt. Vielleicht erbte oder

his thesis design of a theatre a book of other possible layout permutations. He was amused to see them appear, one by one, in other students' schemes. »I had not necessarily taken the best solution,« he said; »but at least I'd been through all the others. Once you know you have exhausted the design potential, it's a very calm atmosphere.«

Steel and timber: material susceptible to logical analysis

But Segal's quest to understand buildings was fruitless until, when a third year student, he read a new book, *Holzhausbau*, by Konrad Wachsmann, Poelzig's pupil of a few years earlier. »It was a revelation.« Here at last seemed something which, amidst a confusion of stylist masters, could be handled comprehensibly and directly. Wachsmann's slim book, cool and finely produced by Wasmuth, has an elegance and poise which is still refreshing. Clearly and simply, it showed the revolution from the framing and joints of traditional wood building to those of the mechanised age, of circular saws and mass-produced nails. It illustrated the American »balloon frame«, »western frame« and »braced frame« construction, and then current German prefabricated systems. Illustrations included Scharoun's S-form prefabricated exhibition house, Wachsmann's fine country house for Albert Einstein, and, hidden away, a beautifully neat, tiny »Sommerhaus« by Poelzig, a clear premonition of Segal's temporary house of 1963!

»I was into a medium I could understand,« said Segal. »This was a rescue and also a delight.« And, with two colleagues, he promptly entered an international competition for the design of a small house. There were 2,000 entries; Segal's won a prize and was published. It had been designed developing the US balloon frame idea, but using a calculated structure, and therefore fewer members. »Since our professors were unfamiliar with the American framing techniques, they had to listen to us!«

Segal only ever discussed this project in relation to structure, but it is a fine exercise in house planning; a prototype L-plan patio house. One can imagine them packing alongside one another into terrace with barely any adjustment, exactly as the patio types which Segal was later to develop. From the criteria of privacy and overlooking, the hierarchy of control of space from the most private to the most public, access from street to back garden, and others which Segal later tabulated, this plan with little amendment could work well.[11]

In Segal's similar later plans (see fig. p. 77 *lower right*), the staircase is pulled towards the front of the house, unlike the dramatic linking element within the main space here. This project makes fascinating comparison with Le Corbusier's »petites maisons«, the two-storey reversed front-to-back terrace houses at Pessac designed five years earlier; while with J J P Oud's terrace at the Weissenhof Siedlung of 1927, the comparison is yet more interesting. (All three – Oud's terrace, Le Corbusier's at Pessac and Segal's competition entry, are designed on the same 5m frontage.)[12]

In the small house competition project by Segal and his colleagues we see a personal voice developing. It is of the moment, but not derivative in form and image. The planning is economical and, as we have seen, has basic common-sense strengths. The structure is economical, its frame is calculated and comprehensibly constructable. It can aggregate to form a pattern of eminently sensible streets in a neighbourhood. But then the surprise: the ground floor, the open 7m by 5m, intersected dramatically by the stair which disappears through the ceiling, hints at a different and more theatrical sense of dwelling.

Revolutionary engineering

»In Berlin I learned one thing, which was the main reason to stay there: there were a number of excellent engineers.« Segal was entranced. »Well, all these things [from ›plastic theory‹ and ›three-moment theory‹ to steel welding] I imbibed as if this was gospel truth. And I determined that every building I was going to make, I would *calculate*.« Segal at last found knowledge, and the foundation of an architectural career in which he never relied on an engineer. »You could see that, from that angle, building was becoming a highly revolutionary, a fascinating process full of possibility.«

An interesting insight into Segal's professional thoroughness comes from a colleague who was gaining practical experience as »Bauführer« after leaving the Technical High School in 1931. The team, where Walter Segal was a member that summer, was working on a hospital building for the Prussian government. The colleague describes how »it was typical practice to put up the carcass of such a building and

lernte er diese Zähigkeit von seinem Vater.« (Julius Posener, Brief an den Autor, 25. 2. 88)

Die systematische Prüfung aller Entwurfsmöglichkeiten wurde für Segal zu seiner lebenslangen Methode. Damals, als Student, trieb er sie ins Extrem. Zusammen mit seiner Diplomarbeit, dem Entwurf für ein Theater, reichte Segal ein Buch mit andern möglichen Grundrißvarianten ein, und er amüsierte sich, als diese dann später, eine nach der andern, in den Lösungen anderer Studenten auftauchten. »Ich hatte nicht unbedingt die beste Lösung gewählt«, sagte er, »aber zumindest hatte ich alle andern durchgeprobt. Wenn man einmal weiß, daß man das Lösungspotential ausgeschöpft hat, gibt das eine ganz ruhige Atmosphäre.«

Stahl und Holz, Materialien, zugänglich für eine logische Analyse

Segals Verlangen, das Bauen ganz zu verstehen, blieb unbefriedigt, bis er, als Student im dritten Jahr, an ein neues Buch herankam: »Holzhausbau« von Konrad Wachsmann, der ein paar Jahre früher Poelzigs Schüler gewesen war. »Es war eine Erleuchtung.« Hier gab es endlich etwas, das durch all die Konfusion der Stile verständlich und direkt hindurchführte.

Wachsmanns schlankes Buch, von Wasmuth zurückhaltend und sauber produziert, hat eine Eleganz und Balance, die immer noch erfrischend sind. Klar und einfach zeigt es die Entwicklung von den Rahmen und Verbindungen des traditionellen Holzbaus bis zu denen des Maschinenzeitalters. Es zeigt die amerikanischen »*balloon frame*«, »*western frame*« und »*braced frame*«-Konstruktionen, sowie die damals gängigen deutschen Vorfabrikationssysteme. Unter den Illustrationen findet sich Scharouns S-förmiges Ausstellungshaus, Wachsmanns feines Landhaus für Albert Einstein und, ganz versteckt, ein wunderschönes, winziges Sommerhaus von Poelzig, das nicht im entferntesten gräßlich aussieht und unverkennbar ein Vorläufer von Segals temporärem *kleinen Haus* von 1963 ist.

»Ich bewegte mich in einem Medium, das ich verstehen konnte«, sagte Segal. »Dies war eine Rettung und auch eine Freude.« Zusammen mit zwei Kollegen ging er sofort an einen Wettbewerbsentwurf für ein kleines Haus. Es gab 2000 Einsendungen; Segal erhielt einen Preis und wurde veröffentlicht. Das Projekt war eine Weiterentwicklung des amerikanischen »balloon frame«-Systems, vereinfachte dieses aber und kam mit weniger Bauteilen aus. »Da unsere Professoren nichts

von den amerikanischen Holzständertechniken wußten, mußten sie uns zuhören.«

Segal sprach über dieses Projekt immer nur im Hinblick auf die Konstruktion, obwohl es auch einen guten Grundriß entwarf, den Prototyp zum L-förmigen Patiohaus. Fast ohne Anpassungen, könnte man sich vorstellen, ließe sich mit ihm eine Reihe bilden, genau wie mit den Patio-Typen, die Segal später entwickelte. Auch Segals weitere Kriterien sind ansatzweise in diesem Grundriß schon umgesetzt: private Zonen, Schutz gegen Einblick, die Hierarchie der Räume vom ganz privaten bis zum öffentlichen, Zugang von der Straße zum Garten.

Das Projekt fordert zu Vergleichen heraus: mit Le Corbusiers »petites maisons«, mit den zweistöckigen, wechselweise nach vorn und hinten orientierten Reihenhäusern in Pessac, fünf Jahre früher entworfen, und, interessanter noch, mit Ouds Reihenhaus in der Weißenhofsiedlung Stuttgart 1927. Diese drei Beispiele haben alle dieselbe Frontbreite von 5m. (Solche Vergleiche wurden seinerzeit in Wasmuths Monatsschriften und, kürzlich, von Philippe Boudon in »Pessac de le Corbusier«, Dunod, Paris, 1969 S. 30, veröffentlicht.) Segal hat später in ähnlichen Grundrissen das Treppenhaus weiter zur Vorderseite gezogen und nicht mehr als dramatisches Trennelement in den Raum gestellt.

Das Wettbewerbsprojekt für das kleine Haus belegt, wie Segal und seine Kollegen eine Denkweise entwickelt haben. Sie entspricht der Zeit. Die Planung ist wirtschaftlich, und zeigt die Stärke alles Vernünftigen. Die Konstruktion ist sparsam und klar verständlich. Man kommt mit diesem Typ zu sehr vernünftigen Lösungen der Reihung, kann sogar ganze Quartiere bauen. Aber dann eine Überraschung: Das ganz offene Erdgeschoß, 5 x 7m, unterbrochen nur durch die Treppe, die nach oben durch die Decke entschwindet; das weist auf eine andere, noch theatralische Auffassung des Wohnens hin.

Revolutionäre Konstruktionen

»In Berlin lernte ich eines, weswegen ich hauptsächlich dort blieb: es gab dort eine Reihe ganz ausgezeichneter Ingenieure.« Segal war hingerissen. »Alle diese Dinge – die plastische Theorie, die Drei-Momente-Theorie, das Schweißen von Stahl – alles saugte ich auf, als sei es die Wahrheit des Evangeliums. Und ich beschloß, jedes Gebäude, das ich künftig planen würde, zu berechnen.« Segal fand hier endlich ein Wissen, das

then when it came to fitting it out, the building was half-destroyed in putting in all the required servicing and equipment. Walter said we must completely design the building, before ever it starts on site, with drawings which show every tap, all the circuits. In fact every hole for every pipe was known before the concrete was cast. It took a long time, but we did it. And just as we finished, the hospital boss was replaced; the new one wanted different equipment and arrangements, and we had to start all over again!

»This time we didn't do it so thoroughly – for by now Walter Segal was not in the team!«[13]

Instead, Segal was using his open scholarship to complete his education at E.T.H. Zurich under Karl Moser[14]. »In Switzerland I learned one very important thing: joinery; I learned it very well indeed. So I was now reasonably well equipped, having some reasonable grounding in building construction, improved very greatly by the knowledge of joinery.

»A fortnight after my exams, a person from Ascona wrote to my mother: would I come and build a small house there? So I went back to Ascona to build, and the 1920s sank back for me into the past. I am still grateful for my decision then, to stay average, in the no-man's land between Bohème and Bourgeoisie.«

Notes and Quotations

1. All this Oedenkoven developed quite on his own, and only noticed in passing by Siegfried Giedion. (*Befreites Wohnen, 85 Bilder erlautert*, Schaubucher 14, Zurich/Leipzig, 1929).

2. *The Architects' Journal*, 7 June 1956, p630.

3. His parents, Arthur Segal (born 1875) and Ernestine Charas (born 1879) were Romanian cousins; at seventeen, she moved with her parents to Berlin, at seventeen he ran *from* his comfortable bourgeois Jewish family to Munich and thence to Berlin. Her family's move was encouraged by growing anti-semitism; his socialist leanings and artistic intentions needed wider horizons than small town provincial Romania could offer. (See the fine catalogue *Arthur Segal 1875-1944*, published by Wolf Herzogenrath and Paven Liska, Argon Verlag, Berlin, 1987.)

4. He asked Freud for a reference when setting up a painting school in England: »Arthur Segal knows the therapeutic and prophylactic value of art, and tries to explain it psychologically; it is to be wished he finds plenty of opportunity to establish himself as a teacher of art.« Sr S Freud, Bergasse 19, Wien 1X, 20 September 1937.

5. »With two others (Julius Posener was one), I collected from forty to fifty drawings, typical drawings made by students in that school. These we published in *Bauwelt*, the standard German professional journal, with an *excessively* aggressive article, you wouldn't believe it! You have to be really twenty-one or twenty-two to write in that sort of way! For Friedrich Paulsen, the editor, this was a really good meal. He gave us eight pages (and the more aggressive parts I contributed). There was a photograph of one of the corridors in the school, full of plaster casts and I wrote the caption: »Here, already in the year 1922, electric light was installed.« When the whole thing came out, you can imagine the impact. There were suggestions that professors should be sacked – it just went like that! The editor was decent and he did not divulge my name.«

6. There is record of his winning third prize in a student competition in 1929, and his second prize project in a 1930 student competition became his first published work: *Bauwelt*, Vol 29, 1930, p992 »Wettbewerbe: Preisausschreiben der T.H.Charlottenburg für eine Ausflugs – Gaststätte – 2 Preis Walter Segal«.

7. particularly by Segal's friend and fellow student under Poelzig, Julius Posener, as in *The Architectural Review*, June 1963, p401(and *Hans Poelzig: Gesammelte Schriften und Werke*, Berlin 1970)

8. Otto Koenigsberger, in conversation with the author, March 1988.

9. Ibid.

10. Julius Posener, letter to the author, 25 February 1988.

11. See Chapter V of *Home and Environment* (1948). In internal arrangement, the layout is almost identical to plan of patio house 1945 (see fig. p. 77 lower right), and is but a slight variation on plans 27 and 24 in that important later book by Segal. The differences all result from the location of the staircase.

12. I have made a comparison of these plans for which there is no room to repeat here; it is not far-fetched that Segal may have compared the Oud and the Corbusier, as such a comparison was published in Wasmuth's *Monatsschriften für Baukunst* at the time; I also refer to a more recent comparison by Philippe Boudon, *Pessac de le Corbusier*, Dunod, Paris, 1969, p30.

13. Otto Koenigsberger, as 8.

14. Though Walter Segal always said this, there seems to be no record of his having enrolled at E.T.H. Zurich, and he did receive a diploma from T.H. Charlottenburg-Berlin in 1932. Could this be part of the mythologising »making of a Swiss«, by which he also encouraged the notion that he had been born in Ascona?

Quotations from Walter Segal in this section from: Wills & Yeo interview 1984; 1973(B)1; 1977(B)2; 1982(C); *The Architects' Journal (AJ)* 21 June 1956, p.701; 1972(B)4; 1969(B); *AJ* 2 June 1970, p.1254; *AJ* 30 August 1956, p.294; 1972(B)1; 1976(C)1; 1976(C)2.

den Grund legte zu einer Karriere als Architekt, in der er niemals auf einen Ingenieur angewiesen war. »Man konnte sehen, daß von diesem Standpunkt aus Bauen zu einem absolut revolutionären, faszinierenden Prozeß wurde, voll von Möglichkeiten.«

Einen interessanten Einblick in Segals professionelle Gründlichkeit vermittelt uns ein Kollege, der als Bauführer praktische Erfahrungen sammelte, nachdem er 1931 die Hochschule verlassen hatte. Das Team, zu dem Walter Segal in jenem Sommer gehörte, arbeitete an einem Krankenhausprojekt für die preußische Regierung. Der Kollege beschreibt: »Es war die typische Praxis jener Zeit, zuerst den Rohbau zu errichten und dann, wenn es zum Innenausbau kam, ihn für all die notwendigen Leitungen und Rohre wieder halb zu zerstören. Walter sagte, wir müßten das Gebäude zunächst fertig entwerfen, ehe wir zu bauen anfingen. In den Plänen müßte jeder Wasserhahn, jeder Stromanschluß zu sehen sein. Wir kannten jeden Durchbruch, für jedes Rohr, bevor der Beton gegossen wurde. Es brauchte sehr lange; aber wir schafften es. Und dann wurde der Chef des Krankenhauses abgelöst, der neue wollte eine andere Ausstattung und an anderen Orten,

und wir mußten wieder von vorne anfangen. Dieses Mal taten wir es nicht mehr so gründlich; denn Walter Segal gehörte nicht mehr zum Team.« (Otto Koenigsberger)

Segal nutzte stattdessen die Freizügigkeit seines Stipendiums und ging an die Eidgenössische Technische Hochschule in Zürich, um dort bei Karl Moser seine Ausbildung zu beenden. Wahrscheinlich ist dies aber eine von Segal selbst verbreitete Legende. Er erhielt sein Diplom 1932 an der Technischen Hochschule Berlin-Charlottenburg.

»In der Schweiz lernte ich etwas sehr wichtiges: Schreinern, und das lernte ich wirklich gut. Damit war ich nun ganz vernünftig vorbereitet, besaß Grundkenntnisse in der Baukonstruktion, die durch die Kenntnisse der Tischlerei noch verbessert worden waren. Zwei Wochen nach meinem Examen schrieb eine Person aus Ascona an meine Mutter, ob ich wohl kommen würde, um dort ein kleines Haus zu bauen. So ging ich zurück nach Ascona, um zu bauen; und die Zwanziger Jahre versanken für mich in der Vergangenheit. Ich bin heute noch dankbar für meine damalige Entscheidung, in der Mitte zu bleiben, im Niemandsland zwischen Bohème und Bourgeoisie.«

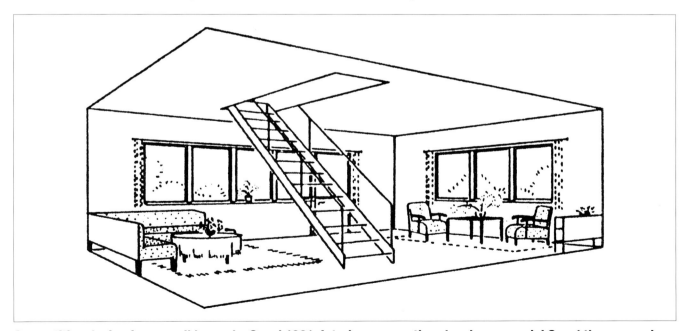

Competition design for a small house by Segal 1931. Interior perspective showing perennial Segal themes such as windows on two sides and a dramatic stair. Published in »Bauwelt«, 1931.
Wettbewerbsentwurf für ein kleines Haus, 1931. Die Innenperspektive zeigt bereits Segal-Dauerthemen wie Fenster nach zwei Seiten und eine dramatische Treppe. Publiziert in »Bauwelt«, Heft 9/1931.

Towards a Non-rhetorical Architecture – La Casa Piccola

A small and most evocative, atmospheric painting by Arthur Segal, dated 1934, shows the view out of La Casa Piccola, his son Walter's first building. By the open doorway, a rush-seated chair has just been vacated, the book left of the edge of the table; the verandah and ballustrade beyond containing the view over Lago Maggiore; a magical evocation. Today it hangs, modestly and easily missed in the pretentious villa in whose grounds hides the little house itself. Arthur Segal's lifelong patron, Bernhard Mayer, had become Walter Segal's first client, commissioning him in 1932 to build this little house, a holiday cabin, in Ascona (see fig. p. 199).

»I was determined to make this my best in performance, design, time and so on,« Walter Segal explained. »By then I *knew* quite a lot about American framing, balloon and platform construction; I knew

Casa Piccola 1987, 55 years after construction.
Casa Piccola 1987, 55 Jahre nach der Erstellung.

Unterwegs zu einer nicht-rhetorischen Architektur

Ein kleines, sehr eindrucksvolles und atmosphärisches Bild von Arthur Segal, 1934 datiert, zeigt den Blick aus der Casa Piccola, dem ersten Bau seines Sohnes Walter. Ein Stuhl mit Binsengeflecht neben der Tür, gerade verlassen, das Buch liegt noch auf der Tischkante. Über die Veranda und das Geländer geht der Blick über den Lago Maggiore, eine magische Beschwörung (vgl. S. 199). Heute hängt das Bild, bescheiden und leicht zu übersehen, in der großartigen Villa, in deren Park das kleine Haus versteckt ist. Es war Bernhard Mayer, der 1932 Walter Segal, über dessen Mutter, aufforderte, dieses kleine Haus zu bauen. »Ich war entschlossen, mein Bestes zu geben, bei der Ausführung, dem Entwurf, der Bauzeit und in jeder Beziehung usw.«, erklärte Walter Segal, der soeben an der TH-Berlin-Charlottenburg sein Architektendiplom erworben hatte.

Casa Piccola shortly after construction in 1932.
Casa Piccola kurz nach Fertigstellung 1932.

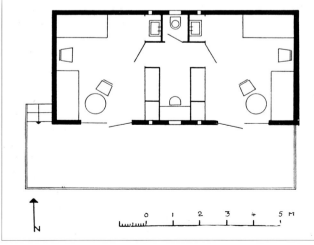

Perspective of proposed cabin and plan (redrawn by Segal for publication in »The Modern House«).
Perspektive des vorgeschlagenen Hauses und Grundriß (nachgezeichnet von Segal für die Veröffentlichung in »The Modern House«).

about joinery. By this time I had changed to freehand drawing which allowed me to draw very quickly indeed. Planning the building to be a timber-framed structure, I produced the design, with all the details and working drawings in pencil, freehand, in four days. It was calculated with a silly thoroughness you wouldn't believe it!«

And it was built in three weeks. Segal claimed to know nothing about foundations, but he luckily knew an old mason in the village (with whose children he had played). »Two carpenters built the balloon frame structure – at first without really understanding. They were

»Damals wußte ich schon ziemlich viel über die amerikanischen Skelettbauweisen, *Balloon*- und *Platform*-Konstruktionen. Auch wußte ich alles über Holzverbindungen. Zu dieser Zeit hatte ich mit Freihandzeichnen angefangen und konnte auf diese Weise sehr schnell arbeiten. Nachdem ich geplant hatte, das Gebäude in Holzständerbauweise zu errichten, produzierte ich den Entwurf mit allen Details und Werkzeichnungen, in Bleistift, freihändig, in vier Tagen. Alles wurde mit so einer blödsinnigen Genauigkeit gerechnet, Ihr würdet es nicht glauben!« (Tonband/84)

Das Haus wurde in drei Wochen gebaut. Segal behauptete, nichts über Fundamente zu wissen; aber »glücklicherweise gab es da im Dorf noch einen alten Maurer (mit dessen Kindern ich gespielt hatte). Zwei wandernde Zimmerleute bauten die *Balloon Frame*-Konstruktion, zunächst ohne sie zu verstehen. Sie waren entsetzt über die geringen Durchmesser. Und sie sagten ›das ist wie Bauen mit Zündhölzern‹. Ich erinnere mich noch daran, und ich antwortete ›das ist amerikanisch‹. Das beeindruckte sie, so daß sie einfach weiterbauten.« (Tonband/84)

»Mein Wissen war in vielen Dingen nur allzu begrenzt: Ich wußte nichts über Toleranzen. Aber diese Leute arbeiteten so fantastisch akkurat, daß das Toleranz-Problem sich nicht stellte. Die Holzverkleidung meiner Wände sollte das Mauerwerk überlappen. Und ich wußte nicht einmal, daß es vernünftig gewesen wäre, hier einen Abstand zu lassen. Ich entwarf es so, daß sich die Bauteile tatsächlich berührten, und so wurde es gemacht. Das Mauerwerk war genau genug ausgeführt.« (Tonband/84)

Was war es, das Segal da so schnell gebaut hatte? »Während vieler Jahre danach«, erinnerte er sich, »habe ich niemals wieder diese Freude empfunden wie bei der Casa Piccola.«

Rings um ihn, in Zürich wie in Berlin – und jetzt auch in Ascona –, zeigte sich die neue, glatte, weiße Architektur: banale Karton-Baukörper, Gebilde aus Chrom- und Glashüllen, die weiche Unverständlichkeit der Betonkonstruktionen. Sein kleines Bauwerk hätte kaum weiter von der Architektur seiner deutschen Lehrmeister entfernt sein können.

Zunächst einmal war es konstruktiv klar und sorgfältig durchgerechnet. Es entspricht Segals lebenslangem Streben nach Verständlichkeit, nach etwas, das man durch Berechnungen vorausbestimmen kann. »Es

gibt keine Komplexität in einer verständlichen, sorgfältig berechneten Schreinerarbeit«, sagte er einmal viel später. Die Casa Piccola beruht auf einer einfachen, soliden Logik. Sorgfältige Berechnung zum Beispiel spart Holz. Und sie nutzt unvoreingenommen die Materialien. Die Casa Piccola war eines der frühesten Häuser, bei denen Holzzement-Platten für das Flachdach verwendet wurden.

Zweitens, untrennbar mit der soliden Logik verbunden, zeigt das Haus eine im Konzept verankerte Klarheit und Offenheit. Alle seine Teile können abgelesen werden. Die Rahmen und Füllungen und Elemente zeigen sich als das, was sie sind. Weder sind sie einem rein visuellen Ordnungssystem untergeordnet, noch werden sie durch irgendetwas getarnt. Fern von einer »brutalen« Überbetonung der rohen Konstruktion, zeigen hier alle Bauteile ihre eigenen Qualitäten, bilden zusammen den Raum.

Colin Ward beschreibt im faszinierenden Buch »Anarchy in Action« die »Produkte, die eine anarchistische Gesellschaft benötigen würde«. Diese verlangt »Objekte, deren Funktion klar ersichtlich ist, Produkte, deren Funktion und Zustand offen liegen«. Dieses Konzept trifft exakt Segals lebenslanges Streben.

Drittens, war das Haus so entworfen, daß es sowohl jene, die es bauten, als auch jene, für die es gebaut wurde, zufriedenstellen sollte. Die enge Beziehung zwischen Bauherr, Architekt und Handwerker war bei der Casa Piccola das Entscheidende und für Segal ein Ziel. Ein halbes Jahrhundert später erinnerte er sich noch gerne an den feinen alten Maurer und mehr noch, an den ausgezeichneten alten Zimmermeister, ein überzeugter Anarchist übrigens – wie auch der Bauherr.

Viertens: Der Plan ist außerordentlich kompakt, folgt einer ganz klaren Typologie. Zwei einseitig belichtete Wohnräume, getrennt durch einen schmalen Streifen mit der Küche nach vorn und Waschgelegenheit nach hinten, absolut ausreichend für diesen winzigen Bau.

Fünftens: Das Haus war billig. Groß war nur der Aufwand an geistiger Energie für den Entwurf. Der Entwurf jedoch war von höchster Genauigkeit. Und das machte es möglich, präzis, ohne Verschwendung von Zeit oder Material zu bauen. Es gab keine Extravaganzen; jedoch (ebenso wichtig) war das Haus auch nicht gewollt spartanisch. Es war eher einfach als asketisch, eher schlicht als puristisch. Es war nicht für den visuellen Eindruck entworfen, jedoch auch

horrified by the small sizes. And they said: ›This is like building with matchsticks!‹ I remember that! So I said: ›It's American.‹ That one impressed them!«

»I knew nothing about tolerances. These people worked to such fantastic accuracy that a tolerance problem did not come up. The timber cladding of my walls was to overhang the masonry. I did not really know that it would be sensible to allow a space! I designed in such a way that it was really touching. And so it was made, the masonry was so accurately built!«

»For many years afterwards,« Segal reminisced, »I never had this enjoyment as at Casa Piccola.« What did this tiny house signify, to remain a paradigm for him throughout his career?

The images all around him, in Zurich as in Berlin – and indeed by now appearing in Ascona – were of the new smooth white architecture. From banal cardboard-like forms via the Bauhaus to the complexity of de Stijl, images of chrome and glass veneers and of the fudging incomprehensibility of concrete design. This little building could scarcely have been further from the architecture of his Germanic tutors.

First, it was structurally clear, and calculated. It met Segal's lifelong goal of realising a comprehensible way of building, which could be understood and thus, through calculation, anticipated. »There is no complexity in comprehensible, calculated joinery,« he said much later. It was based on simple, sound logic (calculation, for example, effecting savings on timber), and on unprejudiced use of material (it was, for example, one of the earliest houses to have a woodwool [»Heraklith«] slab roof).

Second, and of course inseparable from that, it had a conceptual clarity and openness. Its pieces could all be »read«, the frames and panels and elements are themselves, without being either subjected to a purely visual ordering or camouflaged by anything else. Far from a »brutalist« over-emphasis on raw structure, say, here all the elements with their own qualities, make up the place.

In Colin Ward's fascinating book *Anarchy in Action*, the author describes »precisely the kind of product which an anarchist society would need.« This calls for »objects whose functioning is transparent; products having a transparency of operation and repair.« That concept catches precisely an essence of Segal's lifelong aim, and in that sense his work is instinctively libertarian.

Third, it was designed giving satisfaction both to those who made it and to those for whom it was built. This linked relationship, of client, architect, craftsman, was crucial at Casa Piccola and became a lifelong concern of Segal's. Half a century later, he warmly recalled that fine old mason and even more, the excellent old head carpenter.

The other side of this link, the crucial fourth factor, of course, was client satisfaction. Not only were they most happy with it, but indeed their descendants remain so in the 1980s; the occupant proudly shows an unexpected visitor the little building still in first class condition.

Fifth, its planning is extremely compact; a very direct typology of two single-facing dwelling spaces, separated by a narrow service strip with kitchen to the front and washing to the back: absolutely appropriate for this tiny cabin.

Sixth, it was built very cheaply; the only expense was in the concentrated mental energy of designing. Designing, however, included a clear imagining of the construction, which made it possible to build so precisely, with no waste in material or time. There was no extravagance, but (just as important) no imposed spartanness; it was simple rather than ascetic, plain rather than purist. It was not just designed for a visual sensation, nor was it an »Existenzminimum« functional mechanism. Its looseness (and its careful siting) let it breathe.

Talking of this building much later, Segal added: »I felt towards the last of my student years that I had somehow missed many of the things around me. Mine had been a very narrow world of modern architects, painters and some poets. Beyond that was a much wider world that I wanted to explore. I think visual values are fleeting. Buildings seen by the human eye tend to be subjected to foreshortening as far as perspective is concerned. This disturbs pure and clean proportions. The appearance of a building, its texture and colour, are not only dependent on light, sun and rain but also on your mood, and even on what you have eaten. I play this game of fleeting visual values continuously. I am not given to the purist pursuit of proportional precision.«

»To try to hold onto the visual thing that is so very dependent on the operation of our senses and moods, is very difficult. This is why I tried to make this building as insignificant as I could. It was deliberately background architecture, non-conformist in regard to *avant-garde* tenets of the time, and it was anti-ideological.«

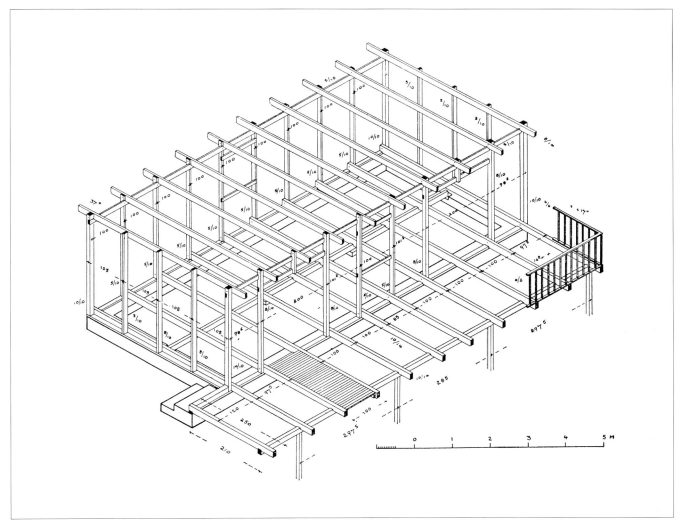

Framing diagram.
Balken-Diagramm.

nicht als »Existenzminimum«. Seine Gelöstheit (und sorgfältige Eingliederung in die Umgebung) lassen es atmen.

Schließlich die Frage, ob der Bauherr zufrieden sein würde. Er war es, sehr sogar, und auch seine Nachkommen, im Jahre 1980, sind es. Der jetzige Besitzer zeigt dem Besucher stolz das kleine Bauwerk, das immer noch in erstklassigem Zustand ist.

Viel später sagte Segal einmal zu diesem Gebäude: »Gegen Ende meiner Studienjahre hatte ich manchmal das Gefühl, etwas von den Dingen um mich herum verpaßt zu haben. Ich hatte in der sehr engen Welt moderner Architekten, einiger Maler und Poeten gelebt. Außerhalb dessen müßte es jedoch noch eine andere Welt geben, die ich entdecken wollte. Ich glaube, visuelle Werte sind fließend. Gebäude, wenn sie durch das menschliche Auge betrachtet werden, erscheinen durch die Perspektive oft leicht verkürzt. Dadurch werden klare, reine Proportionen gestört. Der Eindruck eines Gebäudes, seiner Oberfläche, seiner Farben, hängt nicht nur von

Centrally, here, bringing all these facets together, was the first demonstration of Segal's search for »a natural architecture«.

»By 1932 not only myself but also a number of fellow students were getting tired of almost anything that you listened to in architecture. You were fed up with the traditionalists, you were fed up with the modernists.... I had become very disenchanted with the more orthodox kind of modern architecture – the works of CIAM, and the designs and buildings of the German architects of the time, and I wondered whether I should acquaint myself with other things.

»I was then thinking it might be desirable to produce a ›non-denominational‹ kind of architecture. You can call it ›natural architecture‹. You would design with techniques and with materials quite naturally; you wouldn't have any intellectual aim. In 1932 I had some sort of idea which was taking me away from metals. I wanted to build what I believed in those days was just simply *a natural building*. I had no desire to make a great effort to convince others or to excite others. I didn't want more than to satisfy and particularly to please.

»I wanted to give buildings a relationship, not only to other buildings but also to their surroundings. And not only a visual, but also a physical kind of context, an unobtrusive and unpretentious relationship that gave more than the merely visual things. I also wanted to find a meaning for shapes and this I have found very difficult indeed, to this very day.

»I like buildings that are unobtrusive and not over assertive – I have always been more affected to build buildings that casually fit into an existing environment. for example, building into trees. It is a vulgar misnomer to call this attitude Romantic. But this is a desire to have buildings that are not in a sense self-reliant but are reliant *on* an environment into which they try in one way or another to fit. This environment I have found, unfortunately, is mostly a non man-made environment. I have found it easier to fit these things into some context with trees and meadows and sloping land and so on.

»I like to have the opportunity of making spaces that surround the building, to widen the scope of the inside of the house by a transition from the inside to the outside.«

»I was in a state of rebellion,« he said. »I was determined to make this little building as insignificant as possible – and I passionately wanted it to be liked, which it was. From this plenty can be deduced.« Indeed it can. If it was »innocuous« (his word), that made it no less

precisely intentioned: the Oregon pine plywood wall and ceiling panels, the paper lantern, all warm wood, natural finishes and sunshine. Segal designed the furniture too, taking particular pride in a set of rush seated chairs, a strong polemic in the context he came from. »It was 1932; remember the Bauhaus? This was very different from the interiors I had known. It was warmer, yes, and richer without any doubt. These were instinctive reactions.«

It was Segal's first attempt instead of fashioning architecture, to embody a »coming home« in building. Just as his father's painting of this house captured that arcadian dream, so it was to remain as the image of Walter Segal's root position.

Thus it was understandable that, although being born in Berlin and in fact only living in Ascona for a few impressionable years either side of the age of ten, and then for his first year as an architect, Segal always acted as if the Berlin of his birth and youth was a foreign land, and Ticino his first real home.

»I went back to Ascona to build,« he said. » It became clear to me that one can have a small path and tread it alone.«

Quotations

Quotations from Walter Segal in this section from: 1977(B)2; Wills & Yeo interview 1984; 1976(C)1; 1982(B); 1973(B)2.

La Casa Piccola, Ascona, 1932.

Sonne, Licht und Regen ab, sondern auch von Deiner Stimmung, selbst davon, was Du gegessen hast. Ich spiele dauernd dieses Spiel der fließenden visuellen Werte. Ich halte nichts von dem puristischen Ringen um die Präzision der Proportionen.«

»Der Versuch, sich an die visuellen Dinge zu halten, die nicht in ihrer Präzision zu fassen, die abhängig sind von Deinen Sinnen und Stimmungen, ist sehr schwierig. Deshalb habe ich versucht, dieses kleine Gebäude so unbedeutend zu machen, wie ich nur konnte. Es war gewollt als Hintergrund-Architektur geplant, non-konformistisch im Verhältnis zu den Zielen der Avantgarde jener Zeit, und es war anti-ideologisch.«

Im Grunde ist es, wenn man alle diese Gedanken zusammenfaßt, die erste Demonstration von Segals Suche nach einer »natürlichen Architektur«.

»Damals 1932, hatten nicht nur ich, sondern auch eine ganze Reihe meiner Mit-Studenten genug von fast allem, was man so über Architektur hörte. Man hatte genug von den Traditionalisten und auch von den Modernisten ... Mir mißfiel die eher orthodoxe Seite der modernen Architektur – die Arbeiten der CIAM und die Entwürfe und Bauten der deutschen Architekten jener Zeit – und ich fragte mich, ob ich nicht andere Dinge kennenlernen sollte. Ich dachte damals, ob es nicht wünschenswert wäre, eine Art nicht-sektiererische Architektur zu produzieren. Man kann auch ein anderes Wort dafür brauchen: man kann es ›natürliche Architektur‹ nennen. Man würde mit ganz natürlichen Techniken und Materialien entwerfen; man würde kein intellektuelles Anliegen haben.« (Tonband/84)

»Im Jahre 1932 hatte ich so etwas wie eine Idee, die mich von den Metallen wegführte. Ich wünschte mir, etwas zu bauen, das ich als natürliches Bauen verstand. Ich hatte nicht den Wunsch, große Anstrengungen zu unternehmen, zu überzeugen oder aufzuregen. Ich wollte nicht mehr, als andere zufriedenstellen, und, ganz besonders, erfreuen.

Ich wollte Gebäude in eine Beziehung setzen, nicht nur zu anderen Bauten, sondern auch zu ihrer Umgebung. Und nicht nur visuell, sondern auch in so etwas wie einen physischen Kontext, eine unaufdringliche und anspruchslose Beziehung, die mehr gab als nur das Visuelle. Ich wollte auch den Formen eine Bedeutung geben, das habe ich allerdings bis heute als sehr schwierig empfunden.

Ich mag Gebäude, die unaufdringlich, nicht zu bestimmt sind. Ich war immer geneigt, etwas zu bauen, das sich wie zufällig einer bestehenden Umgebung einfügt, zum Beispiel zwischen Bäume. Es ist ein üblicher Irrtum, so eine Einstellung als romantisch zu bezeichnen. Romantisch-Bauen ist ein sehr komplizierter und widersprüchlicher Vorgang und besteht aus vielen Elementen. Ich hatte den Wunsch, etwas zu bauen, das nicht selbstbezogen ist, sondern sich auf sein Umfeld bezieht und versucht, sich auf die eine oder andere Weise in dieses einzupassen. Ein solches Umfeld habe ich unglücklicherweise vorwiegend in einer nicht vom Menschen geschaffenen Umgebung gefunden. Es war leichter, Dinge in eine Welt aus Bäumen und Wiesen und bewegtem Gelände einzufügen. Ich nutze gerne die Gelegenheit, den Raum, der das Gebäude umgibt, so zu gestalten, daß durch einen fließenden Übergang von innen nach außen der ›Gesichtskreis‹ der Innenräume erweitert wird.«

»Ich befand mich in einem Zustand der Rebellion, ich war entschlossen, dieses kleine Bauwerk so harmlos wie möglich zu machen – und ich wünschte leidenschaftlich, daß man es mögen solle, und das tat man dann auch. Hieraus kann man manches schließen.« Das kann man tatsächlich. Wenn es »harmlos« war, so war dies genau beabsichtigt: Sperrholz-Wand und Deckentäfer aus Oregon-Kiefer, Papier-Laternen, überall warmes Holz, naturbelassene Materialien, Sonnenschein. Segal entwarf auch die Möbel und war besonders stolz auf die Stühle mit den binsengeflochtenen Sitzen, womit er gleichsam gegen sein Herkommen polemisierte. »Es war 1932; erinnert Ihr Euch an das Bauhaus? Dieses war etwas ganz anderes als die Innenräume, die ich gekannt hatte. Es war wärmer, ja, und reicher, zweifellos. Dieses waren instinktive Reaktionen.«

Die Casa Piccola war Segals erster Versuch, Architektur nicht modisch auszustatten, sondern aus einem Gefühl des »Nach-Hause-Kommens« zu gestalten. Ebenso wie seines Vaters Bild von diesem Haus den arkadischen Traum eingefangen hatte, so sollte es Sinnbild für Walter Segals Grundeinstellung bleiben.

So war es denn verständlich, daß Segal, obwohl er in Berlin geboren war und in Ascona nur für ein paar Jahre gelebt hatte, immer so tat, als ob Berlin das Fremde und das Tessin sein erstes wirkliches Zuhause sei. »Ich ging zurück nach Ascona, um zu bauen; es wurde mir klar, daß man einen schmalen Pfad haben und den alleine gehen kann.«

»Building« and Common-Sense

As a teenager, Walter Segal was clear about what he wanted to study: »I wanted to learn to build, first, and perhaps do architecture later if I was able to.« And he held firmly to this line throughout his education. »As a student I didn't really want to discuss with fellow students problems of architecture at all. I never joined that sort of thing, I just dismissed them. And when I was challenged on things, I would say: ›I don't know sufficient about these things. I've got first to know about *building* before I can have opinions on that kind of thing.‹« For him building was the goal, and an uncommon common-sense the method. It was not just that he felt knowledge to derive from empirical study. The springs of his creativity were essentially grounded and rooted in reality; his power exercised through direct knowledge and contact with the sensate world of things.

The essential word is common-sense. Segal's architectural project in this sense is based on a logical step-by-step common-sense more like that of Paxton and Fox at Crystal Palace (1850) or even closer to I K Brunel's, when he designed the timber-frame Crimean War hospital (1855), shipped out to Renkioi and erected so fast – and where Brunel's thoroughness, very like Segal's, even extended to designing details of transportation, unloading and construction by a few chosen carpenters.

The Segal method of building is based on precision technology: total calculation, and optimisation of resources. For a small pickle factory in London, he began by laying a grid of woodwool slabs over the site; he divided it into monitor roof trusses, designed and calculated these steel trusses using mild steel rod diagonal ties (»visually very much less obtrusive«), and there it was.

For his later houses, he calculated frames of finest quality timber. He avoided »second shaping« by careful scouring of the market, and by using uncut, easily available manufactured sheets and panels. Bills of quantities were complete lists of material required on site, with each building element listed separately (not aggregated). Thus, as a whole, it formed an accurate and detailed shopping list. Moreover, because the list was in order of erection, the materials, inevitably but in-

tentionally, could be sorted before delivery. Loaded in this order on the lorry and unloaded in reverse, the materials were stacked on site so that required items were always those nearest to hand, avoiding »second handling«. With each element detailed and located diagramatically, carpenters could use the bill of quantities as a working detail.

As with the common-sense in building, so Segal explored common-sense in design. For clients, his dwellings are full of little details, the most uncommon common-sense, which make everyday life easy. One client slowly noticed many little instances, including: »All light switches were placed at the same height as the door handles so that one never had to search for the switch when entering a darkened room....«

»The staircases had no risers, being composed solely of strings and treads. Each tread was fitted with a square hardwood batten at its leading and rear edges. The battens were secured with brass cups and screws. The width of treads between strings was exactly 27″. This arrangement allowed us to buy a shorter than usual length of standard width stair carpet, cut it into separate pads and secure it with the battens. After approximately five years, when the carpet had worn away along the nosing, I unscrewed the battens and reversed the pads. After a further five years when the carpet nosing had worn again, I cut off the worn part and exposed the

The common-sense stair carpet: Peter Rich's sketch of the treads on his Segal stair.

Der Treppenteppich des gesunden Menschenverstands. Peter Richs Skizze der Handhabung seiner Segal-Treppe.

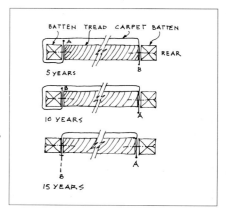

Das Bauen und der gesunde Menschenverstand

Schon als Schuljunge war es für Segal klar, was er werden wollte: »Ich wollte zunächst lernen zu bauen, später dann vielleicht, falls ich dazu fähig wäre, Architektur machen.« Durch seine ganze Ausbildung hindurch hielt er sich an diesen Plan. »Als Student mochte ich die Probleme der Architektur gar nicht mit meinen Kollegen diskutieren. Ich beteiligte mich niemals an diesen Dingen; ich ließ das Thema einfach fallen. Und wenn ich direkt herausgefordert wurde, sagte ich: Ich weiß nicht genug über Architektur. Ich muß zuerst etwas vom Bauen verstehen, ehe ich mir eine Meinung bilden kann.« (Tonband/84)

Für ihn war Bauen das Ziel, und ein ungewöhnliches Verständnis des Gewöhnlichen die Methode. Der Grundbegriff ist der gesunde Menschenverstand. Segals Projekte beruhen auf einem vernünftigen, schrittweisen Vorgehen, etwa wie dem von Paxton und Fox beim Bau des Kristall-Palasts (1850) oder, mehr noch, dem von I.K. Brunel, der ein Hospital in Holzständerbauweise für den Krimkrieg entwarf; es wurde nach Renkioi verschifft und dort ganz schnell aufgebaut (1855). Brunels Gründlichkeit gleicht derjenigen Segals. Sie ging soweit, daß er sogar die Einzelheiten der Verschiffung und der Stapelung der Bauelemente plante.

Segals Methode beruht auf einer präzisen Technik, einer umfassenden Berechnung und Optimierung der Ressourcen. Es geht dabei natürlich nicht um »High-Tech«, sondern um ein System, das als »angemessene Technologie« im Prinzip jedem vernünftigen Menschen zugänglich ist. Beim Bau einer kleinen Konservenfabrik in London begann Segal damit, daß er einen Rost aus Holzwollplatten über den Baugrund legte. Er unterteilte ihn dann in Binderfelder, entwarf und berechnete die Binder, wobei er einfache Stahlstäbe als Diagonalversteifungen vorsah, die »visuell weniger aufdringlich« wirkten, und das war es.

Für seine späteren Häuser verwendete er Holz erster Qualität. Er vermied jedes Nachschneiden, weil er sorgfältig den Fertigteilmarkt erforschte und dann nur leicht erhältliche Paneele und Platten in ihren gegebenen Abmessungen verwendete. Seine Stücklisten sind vollständige Aufstellungen der Materialien, die auf der Baustelle benötigt werden, jedes Element einzeln aufgeführt, nichts zusammengefaßt. Da die Liste darüber hinaus die Materialien in der Folge des Bauablaufes aufführt, können sie vorsortiert, verladen und in umgekehrter Folge wieder abgeladen werden. Das Material wird dann auf der Baustelle so gestapelt, daß immer das, was man braucht, zuoberst liegt. »Doppeltes Hantieren« entfällt. Da jedes Element in den Plänen detailliert und diagrammatisch festgelegt ist, können die Zimmerleute die Stückliste direkt als Arbeitsunterlage benutzen.

Die vollständige Stückliste versieht der Baustoffhändler mit den Preisen. Die Summe ist ein genauer Kostenvoranschlag.

Vom nüchteren Verstand beim Bauen zum gesunden Menschenverstand beim Entwerfen: Für die Bauherrn sind Segals Häuser voll von kleinen Details, die das tägliche Leben erleichtern. Einer seiner Klienten berichtet:

»Der Heizraum hatte ein kleines, nach innen gehendes Fenster ungewöhnlicher Größe, die sogar das Zuschneiden von Backsteinen erforderte. Das Fenster war nach innen angeschlagen. Die äußere Fensterbank, gerade neben der Haustür und etwa 1,5m hoch, war mit Zinkblech abgedeckt, und wir entdeckten bald, daß diese Nische gerade für vier Milchflaschen ausreichte, genau außerhalb der Reichweite der Kinder ...

Alle Lichtschalter lagen in gleicher Höhe wie die Türgriffe, so daß man niemals lange suchen mußte, wenn man einen dunklen Raum betrat.

Die Treppe hatte keine Setzstufen; sie bestand nur aus Wangen und Trittstufen. Jede Stufe war vorne und hinten mit Hartholzleisten versehen, die mit Messing-

clean hardwood nosing. Thus our original stair carpet lasted 20 years!... And so one could continue....«[1]

In Segal's own house similar touches abounded. For example: concerned about acoustic privacy between dining and living room, which were divided by a single-skin brick wall (fair-face both sides), there were two doors, one opening into each room. Moreover, displaying Segal's ability to make common-sense of found products, they close remarkably successfully on door-stops made of square-section bitumenised polyurethane joint filler.

The standard window he developed for his later houses is another typical example. The windows are simply sheets of glass with polished edges and small glass nibs stuck on for handles. With a judicious use of polish, usually a rub with a candle before fixing, they glide in a channel made up of two aluminium angles held apart to allow rain-water to escape.

These are each little examples of the common-sense »method«, most of which apply with ease to little buildings. There are a thousand such radically »simple things« which Segal developed over the years, into a folk wisdom of architecture very similar to that of »Eugène«. In fact, as we shall see, he went much further than Viollet-Le-Duc. For Segal's was not just a system of architectural knowledge; it was integrated into a world-view, exemplified in his understanding of the place of housing, the enabling role of designers, the creative potential of anyone to construct their own dwelling; and it is demonstrated in his own career.

What marked him as extraordinary was his simple, personal need to understand what he was doing. He would not be fooled by other people's bluffing, and he would not bluff himself. This led him to a practice of architectural design which was comprehensible – and not just to himself. One client, with no previous building experience, who built his own Segal house said: »His concept is that if he makes you sit down and think...you will understand what to do.«

This common-sense system is informed by Segal's intuitive concern for personal freedoms and responsibility, which produces an architecture totally under control, quick and pleasant to erect (being dry and with a shell up in a few days), economic in material and labour; enjoyable to design, build and pay for.

Common-sense building uses hand power tools as an appropriate technology, to release the creativity of the individual. Common-sense design springs from cal-

culation. This allows the freedom to make other common-sense decisions.

»Madam how much does your house weigh?« was the startling question Buckminster Fuller was famous for asking in the 1960s. It implied, of course, the lighter the better. It also implied that she should know the answer. By the 1970s, Lloyd Kahn, Bucky protégé (and author of *Domebooks*) just as startlingly recanted: »You don't need to know how heavy your house is,« he realised. »It isn't going to fly.«

But Segal *did* know how heavy his houses were. It was important so that the buildings *didn't* fly. (Wind forces acting on a roof wanting to lift it off are often more than its weight holding it down.) If Segal's common-sense logic shows the advantages of buildings sitting on their foundations without having to be tied down, of roofs sitting on the tops of buildings free to contract and expand without being tied down, then to know their weight is precisely essential (see fig. p. 141 *lower right*).

Only the calculated approach to design makes this common-sense architecture possible.

And he needed to know how heavy each bit was, not just to show calculations of loading and deflection, and to design foundations to hold them, but so that the bits could be conveniently moved, lifted and fixed in place by real, ordinary people – his clients. The important point is that by *the architect* working with calculation and reason and as much precision as possible, the maximum freedon is allowed both (a) to material possibilitites and (b) to human choice and client or builder (which may or may not be the same) involvement.

That tiny timber summer-house by Poelzig which Segal knew from Wachsmann's book, and which surely remained hidden in his memory, had written on the plan: »Gewicht [weight] ca 9300 kg«. The little caption links Poelzig with Wachsmann, Fuller and Segal in this sensationally common-sense concern.

Segal's houses, with their calculated precision technology, are therefore more precisely products of »appropriate technology« than »hi-tech«. If one compares, say, images of the frame of a two-storey Segal house with a typical »late-modernist« two-storey, steel and glass frame, like the award winning house of Patty and Michael Hopkins in north London, the point is quite clear. Each is a remarkably lithe, slender, elegant frame structure; it is a rational, ordered box. But while the

schrauben befestigt waren. Die Breite der Stufen war 77cm. Wir konnten also einen normal breiten Treppenläufer (jedoch ein wesentlich kürzeres Stück) kaufen, diesen in Streifen für die einzelnen Stufen schneiden und mit den Hartholzleisten befestigen. Nach ungefähr fünf Jahren waren die Vorderkanten durchgetreten. Ich schraubte einfach die Leisten los und drehte die Teppichstreifen um. Nach weiteren fünf Jahren, als die Vorderkanten wieder durchgetreten waren, schnitt ich die abgewetzten Teile ab und ließ die sauberen Hartholzkanten sichtbar. Auf diese Weise hielt unser erster Treppenläufer 20 Jahre ...

Und so könnte man weiterfahren ...« (Peter Rich: Notes of a Friendship, A3 Times, PNL, 1985)

In Segals eigenem Haus gab es eine Unmenge derartiger Dinge. Zum Beispiel: Für einen genügenden akustischen Abschluß zwischen Eßzimmer und Wohnraum, die durch eine einschalige, beidseitig unverputzte Backsteinmauer getrennt waren, brachte er eine zweifache Tür an, je eine in jeden Raum zu öffnen. Ein weiteres Beispiel, wie Segal zufällig entdeckte Produkte vernünftig zu verwenden wußte: Die Türen schließen dicht, dank der Schwelle aus einem bituminisierten Polyurethan-Fugenstreifen.

Das Standardfenster, das er für seine späteren Häuser entwickelte, ist ein anderes typisches Beispiel. Die Fenster sind einfache Glasscheiben mit polierten Kanten und kleinen aufgesetzten Glasknöpfen als Griffe. Durch einen geschickten Gebrauch von Gleitmitteln – für gewöhnlich reicht ein kurzes Reiben mit einer Kerze vor dem Einsetzen – gleiten sie einwandfrei in einer Führung aus zwei Aluminium- Winkelschienen, mit genügend Abstand, um das Regenwasser abfließen zu lassen.

A Segal two-storey house, before being clad.
Ein zweigeschossiges Segal-Haus, vor der Auskleidung.

Segal's early explanatory drawing (1966/67) of his timber building method.
Eine von Segals frühen Zeichnungen zur Erläuterung seiner Holzbau-Methode (1966/67).

Es gibt an die Tausend solcher radikal einfachen Dinge, die Segal über die Jahre entwickelte, zu einer Art Volksweisheit der Architektur, sehr ähnlich der des Eugène von Viollet-Le-Duc.

Was ihn leitete, war das einfache persönliche Bedürfnis, stets zu wissen, was man tut. Er ließ sich nie von anderen Leuten etwas vormachen und machte sich selbst auch nichts vor. Dieses führte ihn zu einer Art des Entwerfens, die transparent war, und zwar nicht nur für ihn selbst. Ein Klient, der sich ohne vorhergehende Er-

Hopkins' one is then just wrapped in cling-film, as it were, the Segal one is filled out into architecture.

The contrast is centred in the primacy of the object's image for the Hopkins (and, naturally, for most architects), and its peripheral importance to Segal and his builder-inhabitants. Each building *has* an image, of course. So the Hopkins offer images of their *machine* – structure, skin, and the clarity of technique in their being held together; Segal inevitably lets an image of a *house* enclose his frame, walls seem solid enough, windows are in the right place, the front door, the verandah or balcony are appropriate.

Perhaps another comparison will be illuminating: Charles Eames, who was born in the same year as Segal, and his wife Ray designed their own now famous house in Santa Monica in 1951. It has considerable conceptual similarities with the Segal way of building. Both, for example, aimed to use industrial products as given without an intermediate second forming; and so both have a quasi-industrial image, an order born out of the material and its structural nature. But while Eames purchased standard *components* (for his cladding, for example), Segal used off-the-peg sheet *materials*, saying, »For me there are no ›components‹.« Both show their frames, their varied infil panels and St Andrews' cross bracing. But Segal's crosses, unlike Eames', are designed in the tradition of Norwegian carpentry as compression, not tension members.

And so the elegance of the Eames house, with its bony frame, black and white and full of careful bright colour, cool and pristine as a hi-tech museum, breathes an air different from the more homely, enclosed Segal spaces, the close-spaced intrusive timber rhythm, wooden floor and (on earlier houses) wall covering, enveloping and warm.

For deep cultural reasons, the clean and cool and open imprints itself on the *visual* consciousness of the modern architect (and the RIBA Gold Medal committee which honoured Eames but not Segal) in a way the ordinary, homely, enclosable rooms, the warm and woody image never does. Its importance is every bit the equal of the Eames, but it is essentially of a different nature and not centred on its aesthetic. »I prefer casual visual relations when built,« Segal says, »which do not *show* the conscious effort.... For aesthetic values are fleeting are they not, depending on our mood or the sunshine?...«

Essential differences between Eames and Segal are more explicitly and extremely shown in their design of chairs. Both designed chairs over half a century. In the 1930s, and having been doing genuinely original research on ancient Egyptian chairs, Segal moved to London where his own chair designs were being marketed by Gordon Russell and Heals before the war. Charles Eames, with Eero Saarinen, began designing chairs shortly thereafter.

The Eames achievement is epitomised by the subtle, and luxuriously comfortable lounge chair and ottoman for Billy Wilder of 1956; rosewood ply moulded in three-dimensions, flexible aluminium connectors, composition seating taken to the limits of possibility; all expressed with an elegant clarity – and sold for thousand of pounds. The Segal achievement is epitomised by chairs he made in the 1960s simply because he'd build a new house and needed chairs (see fig. p. 113 *upper right*). Using the latest technology, and with an equally lively imagination, Segal experimented with unjointed timber chairs structurally based on the shear strength of the then new Resin W pva glues. There were no joints, dowels, tenons or mortices. The cantilever chairs, all in beech, were calculated for bending and sheer stresses. He drew force diagrams; the glue manufacturer submitted a report of strength tests. The chairs were cheap, comfortable, practical and gave two decades of good service. They appear ordinary and useful, with a lack of visual bravura which almost blinds one to amazing structural daring, of which Segal's own office chair is a fine example.

Häring and new building

There is no doubt that Segal felt affinity with the design intentions in the group pulled together by Herman Muthesius under the label *Das Englische Haus*. [I remember a fascinating and immensely long lecture he gave about this to my students in 1979.] Even more than to their eminently sensible house plans, or the single-minded attention Philip Webb gave to every detail, Segal responded to their empirical common-sense base, their social building goal and the writings of Lethaby. There are passages in Lethaby, for example on proportion and order, which recur almost word perfect in Segal. (Raymond Unwin, whose absence from Muthesius' book is notable, was also a strong influence whom Segal was pleased to meet on arrival in England.)

But the links between that English »new building« and »Modernism« were too difficult for some. In the late 1920s, as a Berlin student, Segal admitted a by then

fahrung sein eigenes Segal-Haus baute, sagte: »Sein Konzept ist, daß er erreicht, daß Du Dich hinsetzt und nachdenkst, und dann wirst Du begreifen, was zu tun ist.«

Diese Haltung wuchs aus Segals unmittelbarer Sorge um persönliche Freiheit und Verantwortung. Er wollte eine Architektur, die total unter Kontrolle der Vernunft ist, die schnell und angenehm zu errichten ist (in Trockenbauweise, in wenigen Tagen fertig), die wirtschaftlich in Material und Arbeitszeit, erfreulich zu entwerfen ist – und alles zu einem Preis, den man bezahlen kann.

Dieses vernünftige Bauen bedient sich der handelsüblichen elektrischen Handwerkzeuge als einer angemessenen Technologie, um im Individuum Kreativität freizusetzen. Ein vernünftiger Entwurf beruht auf Berechnung. Diese gewährt die Freiheit, weitere vernünftige Entscheidungen zu treffen.

»Madame, wieviel wiegt Ihr Haus?« Das war die berühmte, überraschende Frage, die Buckminster Fuller in den Sechziger Jahren zu stellen pflegte. Die Meinung war natürlich: je leichter, desto besser. Außerdem besagte sie, daß man die Antwort kennen solle. In den Siebziger Jahren konterte Lloyd Kahn, Buckys Protégé: »Sie brauchen nicht zu wissen, wie schwer Ihr Haus ist; es wird nicht fliegen.«

Segal jedoch wußte, wie schwer seine Häuser waren. Es war ihm wichtig, eben damit die Häuser nicht fliegen. Die Windkräfte, die auf ein Dach einwirken und es abzuheben drohen, sind oft größer als das Gewicht. Wenn Segals Logik zeigt, daß es von Vorteil ist, wenn Gebäude fest auf ihren Fundamenten sitzen, ohne daran festgebunden zu sein, daß Dächer besser lose auf die Gebäude gesetzt sind, um sich frei auszudehnen und zusammenzuziehen, dann ist es ebenfalls logisch, daß die Gewichte aller Teile bekannt sein müssen. Weiter mußte Segal wissen, wie schwer jedes Stück ist, um die Teile bequem zu bewegen, zu heben und zu befestigen, und zwar durch gewöhnliche Menschen, seine Klienten, ohne Kran.

Das winzig kleine Sommerhaus von Poelzig, das Segal aus dem Buch von Wachsmann kannte, und das fest in seinem Gedächtnis verankert blieb, trug auf dem Plan die Notiz: »Gewicht etwa 9300kg«. Dieser kleine Schriftzug stellt die Verbindung von Poelzig zu Wachsmann, zu Fuller und Segal her, in ihrem sensationellen Engagement für die Vernunft.

Segals Häuser mit ihrer Präzisionstechnologie sind als »angemessene Technologie«, nicht als »High-Tech« zu bewerten. Man vergleiche zum Beispiel die Holzkonstruktion eines zweigeschossigen Segal-Hauses mit dem Stahl- und Glashaus von Patty und Michael Hopkins. Bei beiden handelt es sich um eine bemerkenswert leichte, schlanke und elegante Skelett-Konstruktion; beidemale eine rationelle, geordnete Schachtel. Die Hopkins aber wickeln sie in Zellophan, Segal gestaltet sie zur Architektur. Der Kontrast liegt darin, daß die Erscheinung bei Hopkins (wie bei den meisten Architekten) Primat hat, während sie bei Segal und seinen Bauherrn nur von peripherem Interesse ist. Trotzdem hat natürlich jedes Gebäude ein Erscheinungsbild. So zeigen die Hopkins-Bauten ihre Maschinen-Struktur und die Technik, die sie zusammenhält. Bei Segal umhüllt die Gesamterscheinung des Hauses die Konstruktion; die Wände scheinen solide genug; die Fenster sind am richtigen Ort; die Haustür, die Veranda oder der Balkon passen.

Segal und Eames

Ein anderer Vergleich: Charles Eames, im gleichen Jahr wie Segal geboren, und seine Frau Ray entwarfen ihr eigenes, inzwischen berühmt gewordenes Haus in Santa Monica 1951. Es zeigt in der Grundauffassung wesentliche Ähnlichkeiten mit Segals Art zu bauen. Beide bemühten sich, Industrieprodukte, so wie sie sind, ohne weitere Bearbeitung zu benutzen. Beide verhalten sich quasi-industriell, beide zeigen eine Ordnung, die auf dem Material und dessen konstruktiven Eigenschaften beruht. Aber die Eleganz des Eames-Hauses, das schlanke Skelett, schwarz und weiß, mit sorgfältig gewählten leuchtenden Farben, kühl und ursprünglich, ein High-Tech-Museum, atmet eine andere Luft als die eher gemütlichen Räume bei Segal, der dichte, eindringliche Rhythmus der Zimmermannskonstruktion, als die hölzernen Fußböden und Wandverkleidungen, umhüllend und warm. Aus kulturbedingten Gründen prägt das Saubere, Kühle und Offene das visuelle Bewußtsein des modernen Architekten in einer Weise, wie es gewöhnliche, Geborgenheit vermittelnde Räume niemals tun. Segals Werk ist Eames' Werk in jeder Beziehung gleichwertig; aber es ist von grundlegend anderer Natur, nicht auf Ästhetik zentriert. »Ich bevorzuge bei Bauten eher zufällige visuelle Beziehungen«, sagt Segal, »die die bewußte Anstrengung nicht so zei-

highly unfashionable regard for the house planning of Mutthesius himself. In the mid 1940s, John Brandon-Jones recalls, »My support for Walter, and my asking him to teach with me at the AA went very much in sympathy with my support of Webb and Lethaby. The common-sense of Lethaby and Walter Segal had very much in common.... But it was most unfashionable. Furneaux Jordan told me that if I didn't stop teaching students all that stuff from Webb and Lethaby, I'd destroy their faith in Modern architecture. So I simply turned round and said ›good afternoon‹, and left the AA.«[2]

Quite differently, threads of that English empiricist approach were taken forward and rewoven during Segal's years in Germany by Hugo Häring into his theory of a »new building« which stood apart from the homogenising International Style. Many of Häring's thoughts have an echo in Segal's work. In *Approaches to Form* (1925) Häring wrote »we should not try to express our own individuality, but rather the individuality of things; their expression should be what they are.« That is, Häring goes on, their function and their material.

We saw how Segal not only knew but plagiarised Adolf Behne's *Der moderne Zweckbau* (1923) before starting architecture school. Behne there says: »By the concept of ›form‹ I do not mean an accessory or an ornament... but rather something deriving from the particular character of the building.«

Both Häring and Hans Scharoun talked of »essential forms« which are particular to each building being designed, and certainly do not derive from either an individual designer's ego or from an imposed aesthetic. It has to be intuited, be teased out. »We must discover things and let them unfold their own forms,« said Häring. And this idealist search for true forms becomes quite metaphysical when he talks of ornamentation as »infringing an object's rights.«

So really »das neue bauen« is a theory of architectural expression rather than of building, of how and why buildings are put together. It is a theory of form and image making, where Häring equated expresion with meaning, before defining it as part of the building's function.

While Segal inevitably knew Häring in Berlin through his parents' circle, much later he claimed absolutely to hate his work. (Intriguingly, some Häring buildings bore a remarkable resemblance to those of Segal, in their natural material and direct handling.)[3]

Segal's ideas of building run parallel to much of »neue bauen« tenets, but clearly they are not part of it. For essentially he is not interested in a theory of expression, but repeats the common-sense adage as if it had no loading on it: »The products of our time exist without the ritualistic symbols of the primitives. A building need not ›express‹ more than it actually is. A book is a book: it has no need in its appearance to offer ›cultural and symbolic‹ expression of its contents. The normal jacket makes no such pretences.«

However, while expression may not be terribly important, that doesn't mean the buildings will just blend innocuously, keeping in keeping with their context. One can be a hard-line moderate, rather than merely displaying peaceful good manners! The »appropriate civilities« approach, typical of the great 19th century engineers, is seen in the behaviour of I K Brunel at Paddington Station. Having the job of designing the station Brunel quite unashamedly and directly asked Matthew Digby Wyatt (a much respected architect) if he would dress it with good manners. The relative roles of »engineering« and »architecture« are crystal clear in the surviving memorandum which Brunel sent his architect.

But there is another response which also accepts that the expression is not of prime importance. This asks: why patronise by being conventional? Why conceal the realities with such sham? The building as instrument may certainly be more important than the building as emblem, but that is all the *more* reason to give it a thoughtful and direct expression.

Segal's constant fight with the forces of conservation and »aesthetic control« was not because his facade design was terribly important. Precisely because it was *not* terribly important it was ridiculous and impertinent to try to control it. His fight was to counter the fixation on the visual. Yet his stance reminds us that Segal had an architect's sensibility, his common-sense was not »just« an engineer's.

Notes and Quotations

1. Peter Rich, who lived in a Segal house (Tasker Road) for 20 years, »Notes of a Friendship«, *A3 Times*, PNL, 1986.
2. John Brandon Jones, in conversation with the author, March 1988.
3. I have this from Peter Blundell Jones, biographer of Scharoun and friend of Segal.

Quotations from Walter Segal in this section from: 1973(B)2; Wills & Yeo interview 1984; 1977(B)2; 1976(C)2; 1969(B).

gen ...; denn ästhetische Werte sind fließend, ist es nicht so? und hängen von unserer Stimmung und dem Sonnenschein ab ...«

Grundlegende Unterschiede zwischen Segal und Eames zeigen sich noch extremer in ihren Entwürfen für Stühle. Beide entwarfen ein halbes Jahrhundert lang Stühle. In den Dreißiger Jahren, nachdem er Forschungen über alte ägyptische Stühle betrieben hatte, zog Segal nach London, wo seine Stühle vor dem Krieg von Gordon Russell und Heals vertrieben wurden. Charles Eames, zusammen mit Eero Saarinen, begann wenig später mit seinen Entwürfen.

Den Erfolg von Eames symbolisieren der sinnreiche und luxuriös bequeme »Eames Chair« und das Sofa für Billy Wilder (1956): Dreidimensional verformtes Sperrholz mit Rosenholzfurnier, flexible Aluminum-Gelenke, die Polsterung perfektioniert bis an die Grenzen des Möglichen, und das Ganze zu kaufen für 3000 Mark. Segals Anstrengungen dagegen sind auf Stühle aus, die einfach, brauchbar und billig sind. Unter Einsatz neuster Technologie und lebhafter Fantasie experimentierte Segal 1960 mit Holzstühlen ohne die üblichen Tischlerverbindungen, sich ganz auf die Stärke der damals neuen Kunstharzleime verlassend. Es gab keine Dübel, Zapfen oder Nuten. Die Buchenstühle waren auf Biegung und Abscherung berechnet. Er zeichnete Belastungsdiagramme, der Leimproduzent führte Versuche durch und lieferte ihm einen Rapport über die Belastungstests. Die Stühle waren billig, bequem, praktisch und leisteten gute Dienste über Jahrzehnte. Sie wirken ganz gewöhnlich, sind ohne visuelle Bravour. Man übersieht dabei fast den erstaunlichen konstruktiven Wagemut, zum Beispiel bei Segals eigenen Bürostühlen.

Segal und das »Neue Bauen«

Es besteht kein Zweifel, daß Segal sich den Gestaltungsgrundsätzen der Gruppe um Hermann Muthesius, unter dem Stichwort »Das Englische Haus«, nahe fühlte. (Ich erinnere mich an eine faszinierende und unendlich lange Vorlesung vor meinen Studenten im Jahre 1979.) Als Student im Berlin der Zwanziger Jahre bekannte sich Segal zu einer damals höchst unmodernen Wertschätzung der Wohnhäuser von Muthesius. Aber auch viele Gedanken Hugo Härings, der das »Neue Bauen« gegen den homogenisierten internationalen Stil abgegrenzt hatte, fanden ein Echo in Segals Arbeit. In

Wege zur Form (1925) schrieb Häring: »Nicht *unsere* Individualität haben wir zu gestalten, sondern die Individualität der Dinge. Ihr Ausdruck sei identisch mit ihnen selbst.«

Sowohl Häring als auch Scharoun sprachen von der essentiellen Gestalt, die jedem Gebäude eigen ist, und keinesfalls dem Ego des individuellen Entwerfers entspringen sollte. »Wir wollen die Dinge aufsuchen und sie ihre eigene Gestalt entfalten lassen«, sagte Häring. »Es widerspricht uns, ihnen eine Form zu geben, sie von außen her zu bestimmen, irgendwelche abgeleiteten Gesetzhaftigkeiten auf sie zu übertragen, ihnen Gewalt anzutun. Wir handelten falsch, als wir sie zum Schauplatz historischer Demonstrationen machten, wir handelten aber ebenso falsch, als wir sie zum Gegenstand unserer individuellen Laune machten.«

Trotzdem ist das »Neue Bauen« eigentlich eine Theorie des Ausdrucks, nicht des Bauens, nicht wie und warum Bauten zusammengesetzt werden. Es ist eine Theorie der Form und der Gestalt, bei der Häring den Ausdruck der Bedeutung gleichsetzt, bevor er sie als eine Funktion des Gebäudes definiert.

Segal muß während seiner Berliner Zeit Häring über den Freundeskreis seiner Eltern gekannt haben. Er erklärte viel später nachdrücklich, ihn zu hassen, obwohl einige von Härings späteren Bauten denen Segals auffallend ähneln, z. B. in der Wahl und direkten Anwendung natürlicher Materialien. Aber obschon Segals Ideen Parallelen zu vielen Anliegen des »Neuen Bauens« zeigen, sind sie doch kein Teil davon. Denn, im innersten Wesen, ist Segal an den Theorien des architektonischen Ausdrucks nicht interessiert. Vielmehr wiederholt er ohne Scheu den nüchternen Ausspruch: »Die Produkte unserer Zeit existieren ohne die rituellen Symbole der Primitiven. Ein Buch ist ein Buch. Es braucht nicht in seinem Äußeren ›kulturell und symbolisch‹ etwas von seinem Inhalt auszudrücken. Der normale Schutzumschlag spiegelt nichts dergleichen vor.« Jedoch, auch wenn der Ausdruck nicht so entsetzlich wichtig ist, sollen die Gebäude nicht einfach – gewissermaßen Ton in Ton – in ihrer Umgebung aufgehen. Das Gebäude als Instrument ist zwar wichtiger als das Gebäude als Wahrzeichen. Aber als Instrument ist ihm ein direktes Aussehen zu geben, soll es sich weder verbergen noch verkleiden.

Chance Moves

The first four or five years of his professional career took Walter Segal from one unpredictable context to another until, equally unexpectedly, he arrived in London and stayed half a century.

»Chance and coincidence are so important, are they not?« For one so light on his feet, it was a marvelous range of experience on which to build a career.

The first surprise was a proposal which arrived from Bruno Taut: »In Spring 1932 Taut suggested that I join his team-to-be in Moscow. I accepted with trepidation, and back in Ascona I struggled with a language which, like the country, met a block in my mind.« But Taut was soon outmanouvred by Social Realism; the project disappeared and Segal remained near Ascona, designing

extensions and » tiny little buildings commensurate with the knowledge which I had,« for a year.

The next surprise was the commission from Dutch friends of Mayer's for a house. »They decided that they could live more cheaply in Palma de Mallorca in Spain. The letter contained a sketch layout of the site and a programme, a very sound effort on my client's part, so I settled down to work.«

And with typical Segal thoroughness; »After a few days of work I counted about 20 schemes or so – all possible, but much too many. They were nicely redrawn and accompanied by little perspectives. In complete foolishness a careful report was prepared weighing the merits of each scheme against the other,

Villa by Segal at Palma de Mallorca, 1934.
Die von Segal gebaute Villa in Palma de Mallorca, 1934.

Ortswechsel

Die ersten vier oder fünf Jahre seines Berufslebens führten Segal von einer unvorhergesehenen Umgebung in die andere, bis er, ebenfalls unvorhergesehen, nach London kam und dort für ein halbes Jahrhundert blieb.

»Zufälle und Zusammentreffen sind so wichtig, meint ihr nicht?« Für jemanden wie ihn, für einen so Beweglichen, war es eine wunderbare Reihe von Erfahrungen.

Die erste Überraschung war ein Angebot von Bruno Taut: »Im Frühjahr 1932 schlug Taut vor, ich solle zu seinem Team nach Moskau kommen. Ich nahm an, mit Zittern, und zurück in Ascona quälte ich mich mit einer Sprache, die, ebenso wie das Land, meinen Geist blockierte.«

Aber Taut wurde vom aufkommenden sozialen Realismus bald ausmanövriert; das Moskauer Projekt löste sich auf, und Segal blieb für ein weiteres Jahr in der Nähe von Ascona, wo er Anbauten und »winzig kleine Häuser entwarf, entsprechend meinen Kenntnissen.«

Mallorca

Die nächste Überraschung war ein Auftrag von holländischen Freunden Bernhard Mayers für ein Haus. »Sie hatten beschlossen, daß sie billiger in Palma de Mallorca leben könnten. Dem Brief lagen eine Planskizze des Geländes und ein Bauprogramm bei, eine sehr ordentliche Leistung seitens meiner Klienten,

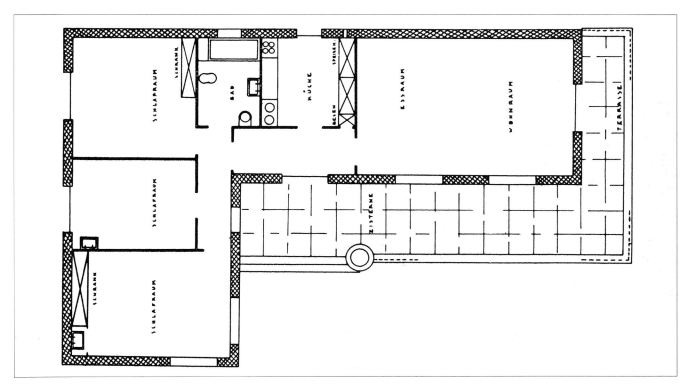

Main plan of Mallorca villa as built 1934.
Grundriß der Villa in Mallorca, 1934.

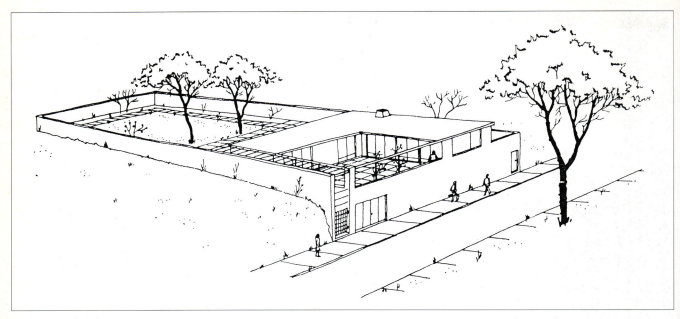

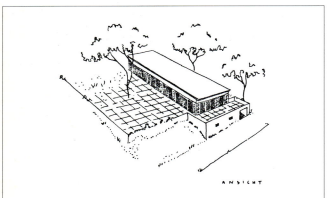

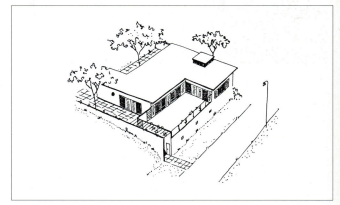

**Various preliminary schemes for the Mallorca villa,
drawn May 1933.**
**Verschiedene Vorentwürfe und Varianten für die Villa
in Mallorca, alle gezeichnet im Mai 1933.**

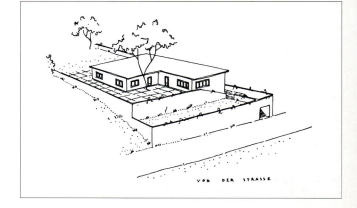

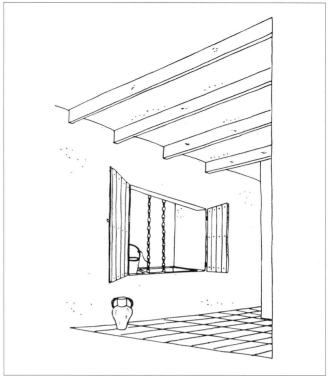

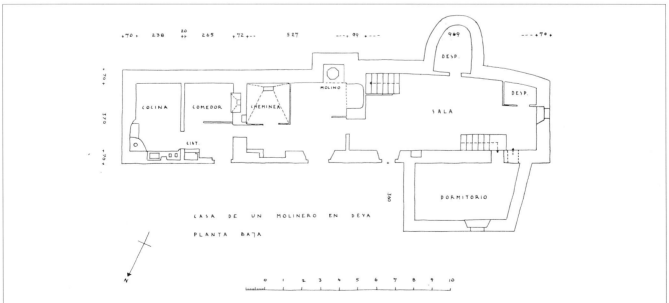

Arthur Segal's painting of Deya, Mallorca (above, left) and Walter Segal's drawings of a miller's house in Deya. Arthur Segals Bild von Deya, Mallorca (links oben) und Walter Segals Zeichnungen des Hauses eines Müllers in Deya.

and the whole lot was sent to Palma and created utter confusion.« The design they finally accepted, when Segal had arrived in Palma, was elegant and of the moment: flat roof, large horizontal windows, the plan a sensible simple L-shape entered at the elbow. But the Modernist forms, inevitably dislocated from local materials and building practice, quickly had to »compromise with Mediterranean practice« (as Segal put it). More interestingly, and more profoundly, his working methods received a shock. Segal's obsession with the total thinking-through of a project before building began, hit complete incomprehension in this insular, traditional but highly-skilled peasant culture. Segal carefully prepared complete drawings and specification and it went to tender to three contractors. Eventually, one builder quoted a lump sum and found little sense in Segal's intention that drawings and specification be part of the contract.

Coming to terms with the traditional contractual and building practices, Segal had an unexpected education. Its model of the delicate equilibrium of a building tradition – balancing its good sense and its skills with its unhasty pace and its conviviality, left a profound memory. Though ambiguously different, it paralleled the equilibrium Segal was developing for himself – balancing his radical, sceptical and reasoned common-sense with the awareness that this must not be allowed to become all-consuming. These builders of Palma worked in a pattern, their periods of intense concentration interspersed with the guiltless periods of release, which so characterises the natural play of a child.

It remained a potent image of unalienated building. But it was not to be envied. »Let us not be romantic. It is obvious that we have to struggle with different problems,« wrote Segal at the time. Much of his subsequent career was spent in working towards technical and contractual solutions which could allow a similarly unalienated building practice for our times and our industrialised culture. »In Mallorca, I enjoyed more than ever before being an architect,« he said later; but it was many more years before he could add: »I've been lucky; I have been able to play.«

Segal designed a few villas in Mallorca and a restaurant on neighbouring Ibiza. He once described that life as an idyll, a lotus-eating existence from which he finally dragged himself away. But he was not idle.

His perennial inquisitiveness produced analyses of the indigenous Balearic architecture. In 1934, he studied the peasant and urban housing on Ibiza, accompanied by Raoul Haussmann, and wrote an extended essay entitled »The domestic architecture of Ibiza«. Then, at Deya on Mallorca, he studied and measured a large number of traditional houses and groups of houses in July the same year, with the help of Heinrich Hossdorf. Perhaps the most evocative image, carefully set in the centre of a white sheet, was a glimpse of houses of Deya in relationship. Finally, it was illustrated with a painting by his father, the centre of Deya framed by a window. This painting Walter Segal kept on his own wall until his death.

The description with which his essay ends help define his own approach to the architecture of dwelling: »The structures resulting from these elements are without pretence. They are simple, and conceived in keeping with the primitive necessities, yet their sense of proportion and clarity cannot be contested.... The Ibicians distrust doctrines where building is concerned,... but they are not content with the bare necessities. It is, with these compositions, not so much a question of ornament as one of proportion, a significant part of all true architecture.

»After a long period of inactivity, the last 20 years had given a new lease of life to European architecture.« But it was now stifled and decomposing, Segal added. The Nazis were already selling postcards of Weissenhofsiedlung as an »Arab casbah« when Segal wrote that.

»It is of particular importance at such a moment, to see a system of architecture where many of our modern ideas are but a very ancient tradition. More than that: composition in which the functional unity of technique and art, to which we aspired, has been achieved. We can learn from Ibiza that this unity is not only the result of pure intuition or good will, but that for its achievement there is the need for an equilibrium of the fundamental conditions of life.«

Egypt

In the paradise of Mallorca, not quite beyond that broadening shadow of Nazism, and with the propinquity of his parents beginning to grate, it was time for the next surprise. He was offered a post as an excavation architect in Egypt, and so it was that he arrived in Cairo, to spend a season at the funerary temple of Thutmosis III[1].

und so machte ich mich an die Arbeit.« Und zwar mit typisch Segalscher Gründlichkeit: »Nach einigen Tagen hatte ich ungefähr 20 Vorentwürfe – alle möglich, aber viel zu viele. Sie wurden sauber neu gezeichnet und durch kleine Perspektiven ergänzt. Als absoluter Narr verfaßte ich einen sorgfältigen Bericht, worin ich die Vorteile eines jeden Vorschlags gegen die jedes andern auswog. Dann wurde das Ganze nach Palma geschickt und führte dort zur absoluten Konfusion.« Der Entwurf, dem die Bauherren endlich zustimmten – Segal war inzwischen in Palma angekommen – war elegant und der Zeit entsprechend: flaches Dach, große horizontale Fenster, ein vernünftiger L-förmiger Grundriß mit dem Eingang am »Ellbogen«. Aber die modernen Formen waren zwangsläufig weit entfernt von den lokalen Materialien und Baupraktiken. Segal hatte bald einen »Kompromiß einzugehen mit der mediterranen Praxis«, wie er sich ausdrückte. Tiefergehend jedoch war, daß seine Arbeitsmethoden einen Schock erlitten. Segals Besessenheit, ein Projekt total durchzudenken, bevor man mit dem Bau beginnt, stieß auf absolutes Unverständnis in dieser insularen, ländlichen, jedoch erfahrenen Kultur.

Segal fertigte sorgfältig vollständige Zeichnungen und den Baubeschrieb an und gab alles an drei verschiedene Unternehmer für die Angebote. Einer nannte einen Festpreis, sah aber wenig Sinn darin, daß Segals Zeichnungen und Materialangaben Teil des Kontraktes bilden sollten. Trotzdem bekam er den Auftrag, und damit begann eine Auseinandersetzung mit den traditionellen Baupraktiken: für Segal eine zusätzliche Ausbildung. Er mußte sich auf ein Gleichgewicht zwischen Vernunft und Erfahrungen einstellen, auf eine gemeinsame Arbeit ohne Hast. Wenn auch aus einem anderen Ursprung, war es doch eine Parallele zu dem Gleichgewicht, das er selber anstrebte. Er mußte seine radikale und skeptische Vernunft mit dem Bewußtsein, daß man dieser nicht erlauben dürfe, alles zu bestimmen, ausbalancieren. Diese Baumeister in Palma arbeiteten nach einem Muster, in dem Perioden äußerster Konzentration mit solchen genüßlicher Entspannung wechselten, wie im Spiel der Kinder.

Was sich ihm einprägte, war das kraftvolle Bild einer nicht-entfremdeten Bauweise. »Laßt uns nicht romantisch sein. Es ist klar, daß wir uns mit anderen Problemen auseinanderzusetzen haben ...« schrieb Segal

damals. Aber in seiner späteren Karriere arbeitete er viel an technischen und vertraglichen Lösungen, die in unserer industriellen Kultur ähnliche, nicht-entfremdete Baumethoden ermöglichen würden. »In Mallorca genoß ich es mehr als je zuvor, ein Architekt zu sein«, sagte er später; aber es vergingen viele Jahre, ehe er hinzufügen konnte: »Ich bin glücklich gewesen. Ich konnte spielen.«

Segal entwarf noch ein paar Villen auf Mallorca und ein Restaurant auf der Nachbarinsel Ibiza. Er beschrieb einmal das Leben dort als Idyll, als »eine lotusessende« Existenz.

Aber er war nicht müßig. In seiner unstillbaren Wißbegierde analysierte er die bodenständige Architektur der Balearen. 1934 studierte er die ländlichen und die städtischen Häuser auf Ibiza und schrieb eine ausführliche Studie über die Architektur des Wohnhauses auf Ibiza. In Deya, auf Mallorca, wo Robert Graves lebte, studierte und vermaß er im Juli desselben Jahres mit Hilfe von Heinrich Hoßdorf eine ganze Reihe traditioneller Häuser und Hausgruppen. Das Bild, das davon am meisten vermittelt, sorgfältig in die Mitte eines weißen Bogens gesetzt, ist eine Ansicht der Häuser In Deya, in ihrer Beziehung zueinander. Noch einmal erschienen die Häuser von Deya in einem Bild seines Vaters, durch das Fenster gerahmt. Dieses Bild hatte Walter Segal bis zu seinem Tode an einer Wand hängen.

Die Beschreibung, mit der er die Balearen-Studie beendet, hilft ihm, seine eigene Einstellung gegenüber der Architektur des Wohnhauses zu definieren: »Die Bauten, die mit diesen Elementen entstehen, sind ohne Schein. Sie sind einfach und im Einklang mit den primitiven Notwendigkeiten, jedoch ihr Sinn für Proportion und Klarheit kann nicht bestritten werden. (...) Die Leute auf Ibiza mißtrauen jeder Doktrin, wenn es das Bauen betrifft, (...) aber sie geben sich nicht zufrieden mit den baren Notwendigkeiten. Es ist, bei diesen Kompositionen, nicht so sehr eine Frage des Ornaments als eine der Proportionen, ein bedeutendes Element aller wahren Architektur.«

»Nach einer langen Periode des Stillstandes hatten die letzten 20 Jahre der europäischen Architektur neuen Lebensgeist eingehaucht; aber der wurde nun erstickt und löste sich auf«, fügte er hinzu. Die Nazis verkauften schon Postkarten von der Weißenhofsiedlung mit der Aufschrift »Arabische Kasbah«, wie Segal schrieb.

He studied ancient Egyptian chairs mainly from the tomb of Tutenkhamen, with the co-operation of Carter who, only a few years earlier, had discovered it. Segal's work was meticulously accurate, thorough and thoughtful. In the words of a modern expert: »It can be said without exaggeration that his records of the Tutenkhamen chairs and stools are invaluable. It goes without saying that it would be impossible to duplicate this work today.«[2]

Though engrossed in the work, Segal's contract was kept to one excavation season by mutual agreement. He »felt that if he'd stayed much more he'd cease to be an architect.«[3] So Segal's next surprise was an invitation to join Erich Mendelsohn, who was setting up practice in Palestine. Segal was not keen, but visited Jerusalem and accepted Mendelsohn's invitation to dinner. Then, with mercurial subtly, he so controlled the evening's conversation that Mendelsohn had no chance to ask him outright, and he avoided having to say »no«. With his voluminous documentation on dynastic chairs, Segal was considering completing a monograph, a fine springboard for an archaeological career. By the winter of 1935 he was drawing and measuring examples in European museums; in Naples, particularly in Turin, and finally in the British Museum.

London

But the next surprise was that the monograph was never completed, and by the end of the summer of 1936 Segal's intensive interest in the ancient Egyptian chairs had receded. He made the decision to become a London architect. He met Eva Bradt, a student at the Architectural Association; their joint entry in a competition for the design of a secondary school in 1937 was commended and published; their practice had begun.

As the final surprise, that year Bruno Taut invited Segal to join him, teaching and designing in Istanbul. Walter Segal had just arranged for the arrival of his parents in England, and had barely settled himself in London – yet another offshore European island, hopefully safe from the continental rumblings; and now he was with Eva. He stayed. »I wanted to be near him, but we could not bring ourselves to go. A year later he died. For me it was more than the loss of the only real link I had with that generation of architects.«

For the next years they lived very simply in London. He designed furniture for Gordon Russell, Heals and Gerald Holtom, and he fitted out domestic bedroom and kitchen interiors. All private building shut down as the War began at the end of 1939, and for the next few years Walter Segal worked with Bertram Carter, designing a number of hostels round the country for workers in factories evacuated from the urban centres, and he designed reinforced concrete air-raid shelters. The boredom of building in the depression and war years was awful, but he now began to write for the architectural magazines[4]. Segal had made the clear decision to be a housing architect. He knew housing would be the central post-war issue in the 1940s England just as it had been in the 1920s Germany of his formative years. So he persued studies of small house design, incorporating continental exemplars and his experience back to student days, in a great range of acceptable solutions whose merits and demerits he mercilessly compared and dissected. Here was a genuine and unique architectural laboratory, bubbling with ideas, conjectures and refutations.

Notes and Quotations

1. I am most grateful to Dr Marianne Eaton Krauss of the Westfälische Wilhelms Universität Seminar für Aegyptologie, and to Prof. Dr Otto Koenigsberger for much assistance on Segal's work at this period, most of which I have been unable to use.
2. Marianne Eaton Krauss, letter to the author, 8 June 1988.
3. Otto Koenigsberger suggested this, in conversation with the author, March 1988. Koenigsberger had been a fellow student with Segal in Berlin; he had obtained a post in Egypt, completed his doctorate on the work, and recommended Segal as his successor.
4. He made contact particularly with William Wood and Noel Musgrave at *The Architect & Building News*, Towndrow and the young Monica Pidgeon at *Architectural Design (and Construction)*, and with W Greenhalgh at *Building*. His major post-war patron was DACA Boyne who had been deeply influenced by Segal as his student in the 1940s. Boyne took the helm at *The Architects' Journal*, guiding it through the 1950s and 1960s to a new goal of professional competence and direct architecture for social purpose, goals which shared much of their flavours with Segal's themes.

Quotations from Walter Segal in this section from: 1976(C)1; 1972(B)1; 1977(B)2; 1942(B); 1934(B)2; 1976(C)2.

»Es ist in einem solchen Moment von besonderer Bedeutung, eine Architektur zu entdecken, in der viele unserer modernen Ideen schon sehr alte Tradition sind. Mehr noch als das: Kompositionen, in denen die funktionelle Einheit von Technik und Kunst, nach der wir streben, schon erreicht ist. Wir können von Ibiza lernen, daß diese Einheit nicht nur das Ergebnis von reiner Intuition oder gutem Willen ist, sondern, daß es dazu auch ein Gleichgewicht in den fundamentalen Lebensbedingungen geben muß.«

Ägypten

Im Paradies von Mallorca, an das der Schatten des Faschismus heranwuchs, in der Nähe der Eltern, was gelegentlich zu Reibereien führte, wurde es Zeit für einen Wechsel. Man bot ihm einen Posten als Ausgrabungsarchitekt in Ägypten an. Und so ging er nach Kairo, um eine Saison am Grabtempel von Thutmosis III. zu verbringen.

Er studierte ägyptische Stühle, hauptsächlich aus dem Grab des Tut-ench-Amon, mit Hilfe von Carter, der es erst vor wenigen Jahren entdeckt hatte. Segals Arbeit war peinlich genau, gründlich und überlegt. In den Worten eines modernen Experten: »Man kann ohne Übertreibung sagen, daß seine Aufzeichnungen über die Tut-ench-Amon-Stühle und -Hocker unschätzbar sind. Es ist ganz selbstverständlich, daß man eine solche Arbeit heutzutage nicht wiederholen könnte.« (Marianne Eaton-Krauss, Brief an den Autor, 8. 6. 1988)

Obwohl er in dieser Arbeit aufging, blieb sein Vertrag, im gegenseitigen Einverständnis, auf eine einzige Ausgrabung beschränkt.

Die nächste Überraschung war eine Einladung, zu Erich Mendelsohn zu kommen, der damals am Anfang seiner Tätigkeit in Palästina stand. Segal war nicht allzu erpicht darauf; aber er besuchte Jerusalem und nahm Mendelsohns Einladung zum Essen an. Es gelang ihm, die Unterhaltung an diesem Abend so zu steuern, daß Mendelsohn keine Chance hatte, ihn direkt zu fragen. Deshalb mußte er auch nicht direkt Nein sagen.

Auf Grund seiner umfangreichen Dokumentation dynastischer Stühle gedachte Segal eine Monographie zu verfassen, als Sprungbrett für eine archäologische Karriere. Über den Winter 1935/36 maß und zeichnete er in europäischen Museen, in Neapel, in Turin und dann im Britischen Museum, London.

London

Aber er stellte die Monographie niemals fertig. Ende des Sommers 1936 hatte Segals intensives Interesse an antiken ägyptischen Stühlen merklich nachgelassen. Er entschied sich, Architekt in London zu werden. »Er fühlte, daß er, wäre er länger bei der Archäologie geblieben, aufgehört hätte, ein Architekt zu sein.« (Otto Königsberger im Gespräch mit dem Autor, März 1988) Er traf Eva Bradt, eine Studentin an der Architectural Association. Ein gemeinsamer Wettbewerbsbeitrag, der Entwurf für eine Sekundarschule, 1937, wurde lobend erwähnt und veröffentlicht. Ihre gemeinsame Tätigkeit hatte begonnen.

Als letzte Überraschung des Jahres lud Bruno Taut Segal ein, zu ihm nach Istanbul zu kommen, zu entwerfen und zu unterrichten. Walter Segal hatte gerade die Ankunft seiner Eltern in London vorbereitet, hatte selbst noch kaum Fuß gefaßt – hier, wieder auf einer europäischen Insel, und, wie man hoffte, sicher vor dem Grollen auf dem Kontinent. Und dann war da Eva. Er blieb. »Ich wäre ihm gerne nahe gewesen; aber wir konnten uns nicht entschließen zu gehen. Ein Jahr später starb er. Für mich bedeutete das mehr als den Verlust des einzigen wirklichen Bindeglieds mit jener Architektengeneration ...«

Über die nächsten Jahre lebten sie sehr einfach in London. Er entwarf Möbel für Gordon Russell, Heals und Gerard Holtom. Er stattete Schlafzimmer und Küchen aus. Alle private Bautätigkeit war zum Stillstand gekommen, als der Krieg begann. Für die nächsten Jahre arbeitete Segal mit Bertram Carter, entwarf eine Reihe von Gaststätten, irgendwo auf dem Lande für die Arbeiter der aus den städtischen Zentren evakuierten Fabriken, und er entwarf Stahlbeton-Luftschutzräume.

Die Langeweile des Bauens während der Depression und der Kriegsjahre war fürchterlich. Er begann, für Architekturzeitschriften zu schreiben. Segal hatte die klare Entscheidung getroffen, ein Wohnbauarchitekt zu werden. Er wußte, daß Wohnungen das zentrale Nachkriegsthema sein würden. So verfolgte er die Studien zum »Kleinen Haus« weiter. Er studierte neue Beispiele vom Kontinent, verglich sie mit seinen Erfahrungen aus den Studententagen. So kam er zu einer großen Sammlung annehmbarer Lösungen, deren Vorzüge und Nachteile er gnadenlos verglich. Das war ein echtes und einzigartiges architektonisches Laboratorium, brodelnd von Ideen, Annahmen und Widerlegungen.

Equipotential Space

The idea of equipotential space, of an openness and loose but precise relation of parts held in equilibrium, is something which was a deep theme throughout Walter Segal's work.

Perhaps he is closest to his father in his search for an order where loose parts are held in dynamic equilibrium and not subservient to the whole. As mentioned earlier, Arthur Segal had developed a theory during the Ascona years of »equivalent value« which he applied to the composition of his paintings as to his politics. Walter Segal aimed for an equilibrium of parts, each with its identity and strength, ordered without hierarchic relationships and only touching gently, and built into casually formed wholes. Such an aim applies to construction techniques, to house planning, or to laying out cities – as well as to the social world.

A »theory of loose parts« like this links well with the notion I quoted from Colin Ward about »transparency of construction and repair«. Each element in an architectural work is seen for itself, yet all are building together.

Such a sensibility implies a working without the preconception of an overall image or external rules such as axiality or an abstract proportioning system. Segal wrote about ancient town forms (in the 1940s): »Flexible regularity offers a balance which we strive to achieve... the free application of geometry. There is no false romanticism in these [classical Greek] towns which had to be laid out as economically as possible. There is no indulging in soft curves for interest's sake... It achieves a balance between the requirements of the individual and the community.

»The hippodamic town is a proof of the thesis that *order is not rigidity; that is the lesson of Ancient Greece.*« [my italics]

At the time he was writing that, Segal also described his own design process in a rather parallel way: »Planning on so limited a scale as in housing is an operation comparable, perhaps, to the playing of a game of chess. In the game, one move determines and restricts the next. Likewise in small house planning....« And he later expanded: »I go step by step, and see how little decisions affect bigger ones; changing the position of a table alters the convenient position of a door, which affects the entrance space, whose alteration means the stair must move, and so it goes on.« The game of chess, with its equally spaced, equal value board is a good image. The game offers a range from directly simple to quite complex moves, whose choice offers an immense freedom of action and range of computable ramifications, but within the utterly precise rules.

As he later said about the way of timber building, »The scope increases if you work within a method. In fact I've far more possibilties now than ever before. I don't know of any other type of building which has this degree of flexibility.... It's quite an astonishing freedom.«

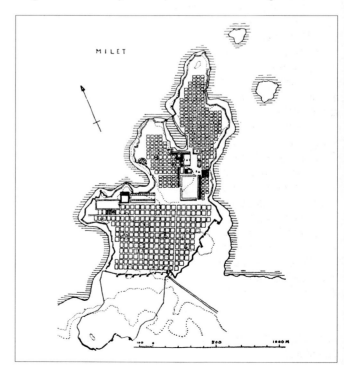

Plan of Miletus drawn by Segal.
Plan von Milet, gezeichnet von Segal.

Gleichgewicht der Räume

Die Idee des Raumes im Gleichgewicht der Kräfte, nämlich einer offenen und losen, jedoch gleichzeitig präzisen Beziehung der einzelnen Teile zueinander, war eines der zentralen Themen, die Segal sein Leben lang beschäftigt haben. Er sucht nach einer Ordnung, in der lose Teile in einem dynamischen Gleichgewicht gehalten, jedoch nicht einem Ganzen untergeordnet werden. Vielleicht kommt er hier seinem Vater am nächsten. Wie schon erwähnt, hatte Arthur Segal in Ascona eine Theorie der »gleich-wertigen Werte« entwickelt, die er sowohl der Komposition seiner Bilder als auch seinen politischen Anschauungen zugrunde legte.

Walter Segal strebte ein Gleichgewicht der Teile an, in dem Sinne, daß jedes Teil seine eigene Identität und Stärke behält, alle Teile jedoch ohne hierarchische Beziehung, sich sanft berührend, ein wie zufällig geformtes Ganzes bilden. Ein solches Vorhaben betrifft Konstruktionstechniken, Grundrisse, den Lageplan einer Stadt oder die sozialen Beziehungen. Eine »Theorie der losen Teile« wie diese paßt gut zu der Vorstellung von Colin Ward, die ich bereits zitierte: Jedes Element eines architektonischen Werkes steht für sich selbst, und doch wirken alle zusammen.

Eine solche Auffassung verlangt ein Arbeiten ohne vorgefaßte Meinungen über Gestalt und Form, über Axialität oder Proportion. »Eine flexible Regelmäßigkeit bildet das Gleichgewicht, nach dem wir streben ... die freie Anwendung der Geometrie ... Es gibt keine falsche Romantik in diesen klassischen griechischen Städten, die so sparsam wie möglich ausgelegt sein mußten. Es gibt kein Schwelgen in sanften Kurven, nur des Interesses halber ... man erreicht ein Gleichgewicht zwischen den Anforderungen des Individuums und der Gemeinschaft. Die hippodamische Stadt ist ein Beweis für die These, daß *Ordnung nicht gleich Starrheit* ist. Das ist es, was uns das alte Griechenland lehrt.«

Zu der Zeit, als Segal das schrieb, formulierte er sein eigenes Vorgehen beim Entwerfen in fast paralleler Weise:

»Wenn man in so kleinem Maßstab plant wie beim Wohnungsbau, ist das eine Operation, die man mit dem Schachspiel vergleichen kann. In diesem Spiel bestimmt jeder Zug den nächsten und schränkt ihn gleichzeitig ein. Ebenso ist es beim Entwerfen von kleinen Häusern.«

Später führte er den Gedanken weiter: »Ich gehe Schritt für Schritt, und sehe, wie kleine Entscheidungen dabei größere beeinflussen. Wenn ich den Platz eines Tisches ändere, ändert das auch die günstigste Position einer Tür, und das wieder beeinflußt den Eingang, dessen Änderung wieder ein Verschieben der Treppe bedingt, und so weiter ...«

Das Schachspiel, mit seinen gleich bemessenen, gleich gewerteten Feldern, ist ein gutes Gleichnis. Das Spiel offeriert eine Reihe einfacher, aber auch recht komplexer Züge. Es gibt eine unendliche Freiheit des Handelns und erlaubt es, Entscheidungen vorauszuberechnen. Aber alles geschieht innerhalb absolut fester Regeln.

Später sagte Segal einmal zur Natur der Zimmermannskonstruktionen: »Der Spielraum wird größer, wenn man innerhalb einer Methode arbeitet. Ich habe tatsächlich mehr Möglichkeiten als je zuvor. Ich weiß nicht, ob irgendeine andere Methode des Bauens diesen Grad an Flexibilität bietet ... Es ist eine ganz erstaunliche Freiheit.« (1976)

Seine Entwurfsmethode folgt Regeln, aber keinem Muster; der Sinn für visuelle Ordnung erlaubt Raum, verhindert jedoch Starrheit. Segal machte viele Studien zu Hausgrundrissen, die innerhalb bestehender Wohnquartiere nicht-hierarchische »Teppiche« von Häusern bilden; hunderte von verschiedenen Studien zum verdichteten Flachbau, zu locker aneinandergefügten Patio-Häusern oder selbst Einzelhäusern (vgl. das nächste Kapitel).

Segal strebte, neben allem anderen, nach Stille, Ausgeglichenheit, Gleichgewicht. Wenig Zeichnungen zeigen das besser als die Perspektive, die er als erste Illustration zu seiner Studie über Deya auf Mallorca anfertigte (Juli 1934). Es ist das eines der wenigen eigentlichen »Bilder«, die Segal während seines Lebens anfertigte. In ihm ist die einfache Fröhlichkeit in einer aus-

His, then, is a method of design which follows rules but not a pattern; a sense of visual order which allows space and not rigidity; and allows for an openness, a looseness in this spacial equilibrium. Segal designed many studies of housing layouts which, within recognisable residential areas, offer non-hierarchical »carpets« of housing; hundreds of varied studies of high-density low-rise housing, loosely touching patio houses or evenly separated individual dwellings, which also (as we shall see in the next chapter) offer considerable advantages over the terraced street or villas round cul-de-sacs of the run of modern house-building.

Segal, apart from anything else, is aiming for a calm, a poise, an equilibrium. Perhaps few images show this better than the perspective he used as his first illustration for the study of Deya in Mallorca (July 1934), one of the very few »pictures« made by Segal in his life. It encapsulates the goal of ordinary conviviality in equipotential space. It has a way of linking buildings loosely together, bound in a topographic context, which is so clearly seen again in his last buildings, »Lewisham 2« (1985). And it closely parallels the remarkable paintings of his father, the dynamic equilibrium of almost hyperreal fruit on a plate (see fig. p. 201).

The loose arrangement of buildings on the ground was a life-long theme. Its traditional model had been seen in the Baleares and its potential realised in one way in Lewisham half a century later. But in the years between, Walter Segal never ceased sketching »equalvalue« housing layouts of more or less formality. They fill his studies of the 1940s (as seen in *Home and Environment* of 1948); again in the 1960s there is a sheaf of studies of high-density low-rise »carpet« housing areas; and in the 1970s he developed a »twenty house to the acre« project into the solution for a real client. This, houses for Solon Housing Association, was refused building permission on visual grounds.[1]

Many other careful design factors support such a scheme, of course; but undoubtedly the sensibility which I describe as »equipotential space« is the germ from which it grows.

The houses of Deya, Mallorca, in dynamic equilibrium in space, as seen by Walter Segal, 1934.
Die Häuser von Deya, Mallorca. Dynamisches Gleichgewicht der Räume nach der Vorstellung von Walter Segal, 1934.

Walter Segal's unassuming and undifferentiated terrace facade from 1944.
Segals zurückhaltende und ruhige Reihenhausfassade von 1944.

gewogenen Umgebung eingefangen – hier wurden Gebäude lose miteinander verbunden, eingebettet in einen topografischen Kontext, wie wir ihn dann wieder so klar in seinen letzten Bauten, in »Lewisham 2« (1985) finden. Hier zeigt sich auch eine enge Parallele zu einem bemerkenswerten Bild seines Vaters, das fast hypernaturalistisch Früchte auf einem Teller darbietet (vgl. S. 201).

Sein Leben lang hat Segal nie aufgehört, Skizzen zur Gruppierung gleich-wertiger Häuser anzufertigen. Sie füllen seine Skizzenbücher der Vierziger Jahre (zu finden in *Home and Environment*, 1948); in den Sechzigern folgt dann ein ganzes Bündel von Studien zu Teppichsiedlungen im verdichteten Flachbau, und in den Siebzigern entwarf er ein Projekt »30 Häuser pro acre« (1 acre = 4047m²) für einen Kunden. Diesem, Solon Housing Associates, wurde aus visuellen Gründen die Baugenehmigung verweigert.

Viele von Segals Wohnbauprojekten sind leider nicht gebaut worden, nachdem 1950 die Nachkriegs-Standards für den Wohnungsbau durchgepreßt wurden und Segal sich weigerte, solch unzulängliche Häuser, wie sie die Standards verlangten, zu bauen. Es war besonders tragisch, daß das Projekt »30 Häuser pro acre«, das zweistöckige Häuser in Holzständerbauweise für arme Leute vorsah und alle anderen rechtlichen und finanziellen Hürden genommen hat, 1978 durch die Vorurteile lokaler Bürokraten blockiert wurde.

Segal war immer gegen Hast. Obwohl er (wie einer seiner frühen Mitarbeiter ihn beschreibt) »hitzig, erregbar war und mit einer haarsträubenden Geschwindigkeit arbeitete«, (Philip Powell im Gespräch mit dem Autor 1988), war sein Ziel eine Art gebauter Stille. Hierauf beruhte auch zum Teil seine Abscheu vor dem Expressionismus, ebenso wie vor romantischen neoexpressionistischen Ideen z. B. von Archigram, deren Projekte er treffend beschrieb als »Wohnungen, die zu einer Kapsel geschrumpft und dabei in eine überspannte Bewegung geraten sind« (The Architects' Journal 22. 6. 1966/S. 1510). Auch am Southbank-Komplex in London mit der Hayward Gallerie und der Queen Elizabeth Hall kritisierte er die »Stimmung von Ruhelosigkeit und Unwirklichkeit«.

Das Überspannte verstand er als Folge einer Über-

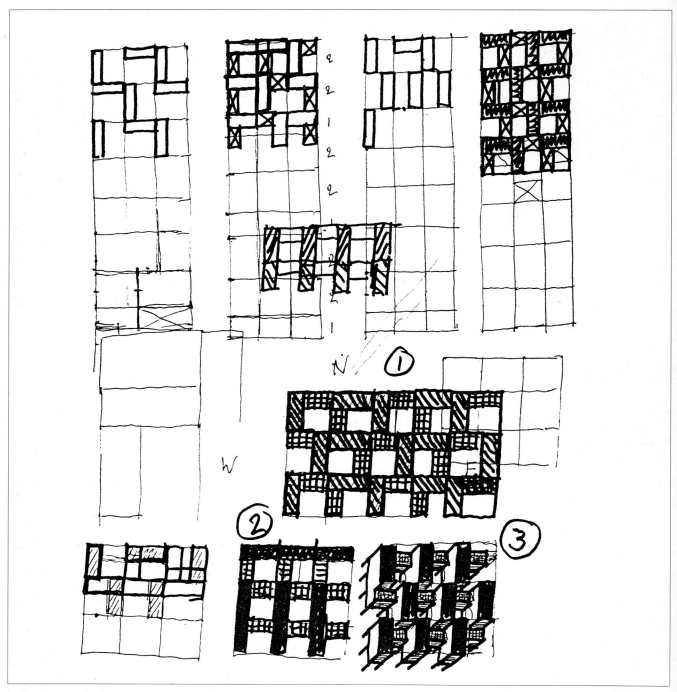

Pages from many A4 sheets of »chessboard« studies of »carpet« housing patterns, and of dense blocks of patio houses, 1971.
Einige von vielen A4-Seiten mit Skizzen von Schachbrettmustern für Teppichsiedlungen und verdichtete Patiohäuser, 1971.

Lewisham 2 facade detail, 1985.
Fassadendetail in Lewisham 2, 1985.

bewertung des Visuellen. Einer der wenigen Artikel, die er von seiner Zeit in Kairo bis zu seinem Tode aufbewahrte, behandelte diese Dinge mit viel Feingefühl. Er ist von Bruno Taut und geschrieben in Japan, angesichts des Katsura Palastes.

Taut stellte diesen berühmten kleinen kaiserlichen Palast einem anderen, der gleichzeitig unter stark dekorativem chinesischen Einfluß in Nikko entstanden war, gegenüber.

Segal war natürlich klar, daß gewöhnliche Wohnarchitektur nicht mit Palastarchitektur zu vergleichen ist. Aber auch seine Häuser ermutigen zu ruhiger Betrachtung, verlangen das »denkende Auge«.

»Ich würde niemals meine Klienten und ihr Geld als Mittel zur Selbstdarstellung mißbrauchen. Die Lösungen sollten immer optimal und typisch sein, nicht individuell. Das Resultat, das man so erzielt, ist selten umwerfend. Je besser ein Entwurf funktioniert, um so weniger ausgeprägt wird er erscheinen, um so weniger wird er sich zur Geltung bringen.« (1968)

Viele Jahre früher, als er eine Besprechung über ein Buch mit Neuen Schweizer Häusern schrieb, klingt es, als beschriebe er sein eigenes Werk: »Was den architektonischen Ausdruck betrifft, so zielt die Tendenz immer auf ein ruhiges Wohlbefinden, und das mag vielleicht jene enttäuschen, für die scharfe Kontraste und dramatische Akzente zur guten Architektur gehören, die solches selbst im Wohnbau erwarten und an wirklichen Leistungen vorbeigehen, ohne sie zu bemerken. Oberflächlich gesehen sind diese Häuser bescheiden und einfach; aber die Detaillierung ist superb. Einfache Architektur richtig zu würdigen, braucht ein gutes und wohlgeschultes Auge.« (The Architect and Building News, 9. 3. 53, S. 437)

Walter Segal was always against frenzy. Though (as one of his early assistants described) Segal was »fiery, excitable and worked at hair-raising speed«[2], the goal was a certain imposed stillness. This sense was part of his distaste for expressionism, as for romantic neo-expressionist ideas (like those of Archigram one of whose schemes he accurately described as »dwelling shrinking to a capsule, and movement over-stimulated«, and for similar neo-expressionist buildings (like the »South Bank« complex in London) characterised by »a mood of unrest and unreality.« Part of this over-stimulation he saw as an excessive emphasis on the visual. One of the very few articles he kept, from his time in Cairo up to his death, sensitively engaged these issues. It is by Bruno Taut, writing while in Japan of the Katsura Detached Palace.[3]

Taut first contrasted that famous, small imperial palace with another, built at the same time under highly decorative Chinese influence at Nikko.

Of course Segal understood that ordinary domestic architecture is not a palace built without budget limitation, and comparison with his houses would be absurd. But they too encourage quiet observation, the thinking eye.

»I never use my clients and their resources as a means of self-expression. Solutions should be optimal and typical, not individual. A result thus obtained is rarely startling. The better a design performs, the less pronounced will it appear, the less assertive.« Many years earlier, he could have been describing his own work when he reviewed a book of new Swiss houses: »As to the architectural expression the tendency has always been for quiet well-being and this may, perhaps, be a source of disappointment to those who associate strong contrast and dramatic accentuation with good architecture; they even expect it in domestic work and are wont to sidepass real achievements without noticing them. Superficially these dwellings are modest and simple, but their detailing is superb. Real appreciation of simple architecture depends on a good and well-trained eye.«

»Modern man,« as the text by Taut notes, »here encounters a particular difficulty: we are accustomed to regard architecture as image, and for the sake of the image effect.« Misunderstood classical architecture and the tyranny of symmetry lead to the »facade«, in German significantly called *Schauseite*, »showside«, differentiated between front and back; all things and ideas which find no place in Katsura.

»Is it not indeed something very simple – and something so self-understood and therefore beautiful – when the entrance, the house and the spaces, garden and everything don't present themselves as soldiers, in regimented formation to left and right of the centre, but instead, each individual piece is placed precisely according to its purpose and its good sense, like a living being? A proper society of free individuals. A true miracle in our world, this arrangement of house and garden which is nothing else than a tight representation of relationships....«

»It is not the artistic ingenuity which makes this building a wonder of the world: it is its absolutely modern approach; it is its style of relationships; that is to say, its built relativity....«

We could compare the precise but loose arrangement of a Segal house plan, such as the Halstead House built on various levels round existing trees in 1969, alongside the Katsura Palace and Taut's comments just quoted.

»I prefer casual visual relations when built, which do not *show* the conscious effort. For aesthetic values are fleeting are they not, depending on our mood or the sunshine?«

Segal argued against the seductiveness of images. From this base, he continually scorned the romanticism of picture makers – whether in town-planning or building design, as well as »the confused romanticism« of polemicists – from Walter Gropius' eulogising craftsmen (»in prose of the purest Morris«) to those in the 1980s icing their architecture with symbolism.

Segal saw »modern« architecture in terms of the international consensus by the mid 1920s, whose essence he described: »The buildings of the International Style were by definition – and this is an essential aspect – unassuming. They avoided drama and ostentation. ... They were meant to promote well-being, e.g. in visual terms to form a quiet background for people, to be un-assertive, unobtrusive, receding – not negative qualities these. Above all they were socially and communally oriented at least by intent and definition; they possessed enough individuality to be recognisable but able to fit into a wider context, an environment. It became possible to let one's awareness of them sink into one's subconscious, the true quality of architecture that has an everyday meaning.«

This outlines the position Segal held to, and whose

Compayne Gardens garden elevation, 1965.
Gartenseite des kleinen Wohnblocks in Compayne Gardens, 1965.

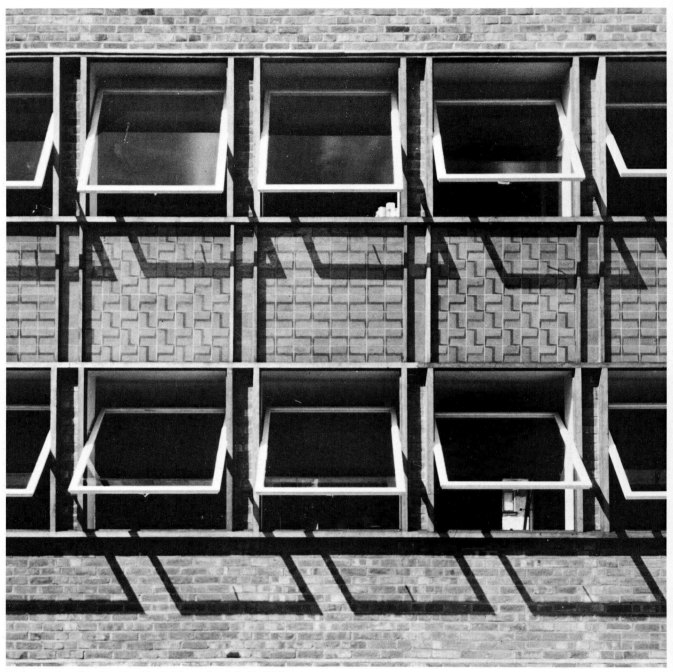

Balanced patterns, the Tretol facade detail, 1954.
Ausgewogene Muster. Detail der Tretol-Fassade, 1954.

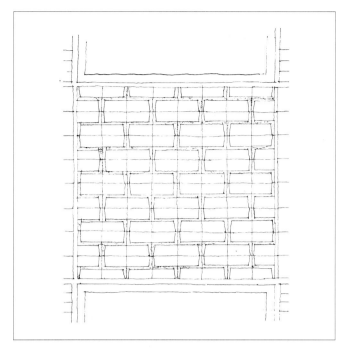

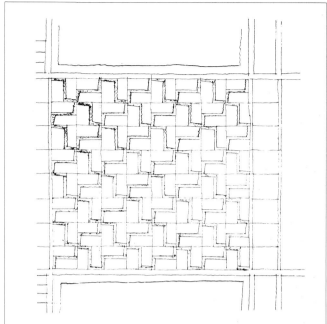

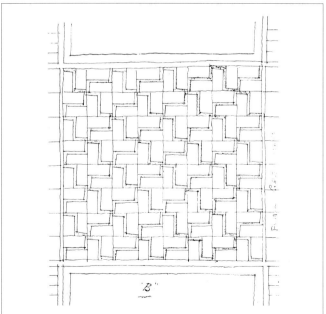

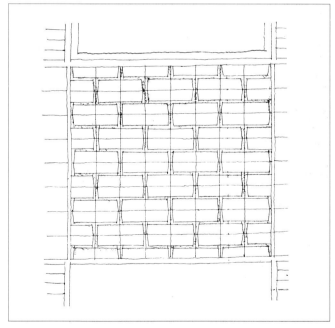

Various proposed arrangements of tile patterns on Tretol facade.

Verschiedene Entwürfe für die Ziegel-Musterung an der Tretol-Fassade.

rightness seemed to strengthen rather than fade (as does a fashion) through his long career.

»Unassuming and lacking in ostentation... the essence of the International Style,« is how he described the architecture of J J P Oud, whose friendship he had valued. There is a concern to play down »expression« and the visual impact, but of course it cannot help becoming an aesthetic concern itself. The elevation of Oud's terrace at Weissenhof has been called »undifferentiated«[4], and this is a word which Segal might indeed have taken up. There is certainly a link, right through to Segal's later buildings, with this aesthetic from the 1920s. In a similar vein, in 1918 Poulenc gave music the marking »balancé – modéré« and »très modéré – indifférent«[5]; it was calling for a certain even, unaccentuated, unhierarchical ordering. »Eschewing special effects,« as Segal said.

Segal's father, shortly before his death, described[6] how (in Ascona in 1916) he had suddenly realised that, in art as in life, there need not be dominating and subordinate elements; that in the social world the urge to dominate was the chief cause of conflict; and that »nature gives the same attention to a butterfly as she gives to a star.... In nature everything is of equal importance and interest.« Thus, he said, his »optical equivalence«, his »equi-balance« was created.

Later (at the end of the 1920s), in contrast both to the Impressionism of his youth and to the later Expressionism, he sought a stillness, »the static element, a concentration on the unchanging nature of the object,« and this too filled his son's work. »A still timelessness, something to hold on to, a rest,« as Arthur Segal concluded.

Walter Segal was fascinated by even-patterning and never tired of playing with such forms – on the scale of town planning, housing neighbourhood layouts, building facades, and even tile patterns and tile design.

Some of his facades – like the small block of flats at Compayne Gardens (1965) or the masterly Ovington Square flats (1964), display this care to a remarkable degree. Ovington Square, where a hardwood frame is layered as a pattern over the blue brick and glass facade, produces a remarkably fine note of coolness which makes the surrounding stucco terraces seem pretentious and overweight. That the *Architectural Review* called this facade »a text-book of bad manners«[7], I will return to. His conscious efforts to avoid the visually seductive, whether self-expressive or camouflaged as keeping-in-keeping, didn't always win friends.

At the Tretol factory (1955) he had used a similar form of window, framing a grid in dark hardwood which united the two floors with an infil of red and grey tile patterns. There are many sketches for these tiles; in the end he chose one, Z-shaped. There are many sketches of how these could combine to make different patterns; in the end he chose to use just two, alternating, on this wall.[8]

The motif of even, undifferentiated forms is quiet; it has to be listened for with care. Segal was modest: »I tried to find my own way, easily accepting my limitations and the fact that I was not going to leave a fantastic mark on a visually-oriented people.... Practical ability does not necessarily produce good buildings; visual inventiveness is fleeting and fickle....To convince, a shape needs more than proportions and texture or colour. Acceptance by the mind is essential for a building to be convincing.«

Notes and Quotations

1. So many of Segal's housing proposals seem not to have been built, particularly after post-war housing standards were squeezed in 1950, because of his refusal to compromise substandard dwellings which the government bureaucracy would pay for (the »housing cost yardstick« as it later became known for many years in the UK). It is therefore tragic that this project, a scheme of two-storey timber-frame houses for poor people, which stood to clear all the other financial and bureaucratic hurdles, was finally stopped in 1978 by the prejudice of one local government bureaucracy.
2. Philip Powell, in conversation with the author, April 1988.
3. Bruno Taut, »Das Architectonische Weltwunder Japans«, *Nippon 11*, 1935, pp2–4.
4. Philippe Boudon, *Pessac de le Corbusier*.
5. Francis Poulenc, the markings of two of his *Mouvements Perpetuels*, J&W Chester, London, 1919.
6. Arthur Segal *Short Autobiography* dated 1936/1943, published in catalogue of Arthur Segal Memorial Exhibition, 18.9.1945 – 1.10.1945, Royal Society of British Artists Galleries, London SW1.
7. Ian Nairn, *The Architectural Review*, January 1959, p72.
8. »I decided to design it in the faint hope that both the avant-garde and Victorians would not approve. The Victorians have done so already,« he said of the Tretol tiles facade; *The Architect and Building News*, April 1955, pp132–7.

Quotations from Walter Segal in this section from: 1948(B); 1976(B)2; 1945(B)1; *AJ* 22 June 1966, p.1510; 1972(B)4; 1969(B); 1976(C)2; 1972(B)3.

Wir könnten die präzise, jedoch lockere Komposition eines Segal-Grundrisses, zum Beispiel den des Halstead-Hauses, das 1969 auf verschiedenen Ebenen um einen vorhandenen Baum herumgebaut wurde, mit dem des Katsura-Palastes vergleichen. »Ich ziehe zufällige visuelle Beziehungen vor, die, einmal gebaut, die bewußten Bemühungen nicht zeigen; denn ästhetische Werte sind fließend, sie hängen von unserer Stimmung oder dem Sonnenschein ab, ist es nicht so?« (1976)

Segal argumentierte gegen die Verführung bildlicher Eindrücke. Hierauf beruhte auch seine Verachtung des Romantischen, der »Bildermacher« – sei es in der Stadtplanung oder in der Architektur oder im Design. In den Achtziger Jahren ärgerte er sich über »Architektur mit Symbolik«, also über die sog. Postmoderne. Segals Urteil über die »Moderne«, die Architektur des internationalen Konsenses Mitte der Zwanziger Jahre, zeigt seine Position: »Die Bauten des Internationalen Stils waren ganz ausdrücklich – und das ist ein essentieller Aspekt – nicht anmaßend. Man vermied alles Dramatische und jedes Gepränge. (...) Sie waren bestimmt, Wohlbehagen zu verbreiten, und das heißt im Visuellen, einen ruhigen Hintergrund für Menschen zu bilden, nicht bestimmt, nicht zudringlich, zurückhaltend – und das waren keine negativen Qualitäten. Darüber hinaus waren sie auch sozial, auf die Gemeinschaft bezogen, zumindest in ihrer Absicht und Definition; sie besaßen genug Individualität, auch in einem weiteren Kontext, einem Umfeld erkennbar zu bleiben. Es war möglich, sie wahrzunehmen, in das Unterbewußtsein einsinken zu lassen, wahre Qualität der Architektur, die Tag für Tag etwas bedeutet.« (The Architect and Building News 25. 9. 1969/S. 27)

»Nicht bestimmend und ohne Gepränge ... die Essenz des Internationalen Stils«, so beschrieb er die Architektur von J.P. Oud, dessen Freundschaft er hoch geschätzt hat (1969). Die Ansichten von Ouds Reihenhäusern in der Weißenhofsiedlung wurden von Philippe Boudon als undifferenziert bezeichnet, was Segal positiv aufgefaßt hätte. Er war bestrebt, den visuellen Eindruck herunterzuspielen, und das wurde unwillkürlich zu einem ästhetischen Anliegen. Diese Ästhetik der Zwanziger Jahre wirkte bis zu Segals letzten Bauten. In einem ähnlichen Geist ordnete Poulenc 1918 die Begriffe: »ausgeglichen – gemäßigt und sehr gemäßigt – indifferent«. »Jeden speziellen Effekt scheuend«, wie Segal sagte.

Segal war fasziniert von gleichmäßigen Musterungen und niemals müde, mit solchen Formen zu spielen – im Maßstab der Stadtplanung, der Quartierpläne, der Fassaden und selbst bei der Verlegung von Fliesen.

Einige seiner Fassaden, wie bei dem kleinen Wohnblock *Compayne Gardens* (1965) oder dem meisterhaften Wohngebäude am *Ovington Square*, zeigen das in hohem Maße. Am *Ovington Square* bewirkt der Hartholz-Raster, der wie ein Muster über die Fassade aus Glas und blauem Backstein gelegt ist, einen feinen Hauch von Kühle, der die benachbarten Reihen mit ihren Stuckverzierungen anmaßend und übergewichtig erscheinen läßt.

Darauf, daß die Architectural Review diese Fassade als ein »Textbuch schlechter Manieren« (Nr. 1/1959/S. 72) bezeichnete, werde ich zurückkommen. Seine bewußten Anstrengungen, der visuellen Versuchung zu widerstehen, brachten ihm nicht immer Freunde.

Bei der Tretol-Fabrik (1955) benutzte er eine ähnliche Fensterform in einem dunklen Hartholzraster, der die beiden Geschosse mit ihren Füllungen aus roten und grauen Ziegeln zusammenfaßte. Es gibt viele Skizzen, wie man diese zu verschiedenen Mustern hätte verlegen können. Zum Schluß wählte er eine Z-Form. Zu dieser Fassade schrieb Segal: »Ich hatte mich dazu entschlossen in der Hoffnung, daß sie sowohl der Avantgarde wie den Viktorianern mißfallen möge. Die Viktorianer haben schon reagiert.« (1955)

Das Thema der ebenmäßigen, undifferenzierten Form ist Ruhe; man muß sorgfältig lauschen. Segal war bescheiden: »Ich versuchte meinen eigenen Weg zu finden, meine Grenzen zu akzeptieren und die Tatsache, daß ich nicht einen phantastischen Eindruck bei einer Design-orientierten Gesellschaft hinterlassen würde.« (...)

»Soll ein Gebäude überzeugen, so ist es unerläßlich, daß es durch den Geist akzeptiert wird ... Wir müssen Aspekte studieren, die es über den Bereich des praktischen Nutzens und des ansprechenden Aussehens emporheben ... Der heutige Formgeber ist einem wirklichen Verständnis für den intellektuellen Prozeß, der notwendig ist, seine Formgebung zu ermöglichen, noch nicht näher ... Praktische Fähigkeiten führen nicht notwendig zu guten Bauten; visueller Einfallsreichtum ist fließend und launisch ... Um zu überzeugen, braucht eine Form mehr als Proportionen, Textur und Farbe.« (1977)

»The Housing Larder«

Housing research

In the absence of real opportunities to build, Walter Segal's major architectural contribution in the dark decade from the late 1930s to the late 1940s was »research ... and the study of design problems without the limitations imposed by the requirements of a client.« Segal described this work as stocking up his architectural larder, and he soon began to open it to the public when, from 1942, he published the studies of housing which eventually were collected together in April 1947 for the book *Home and Environment* (published 1948).

Seen together, this was a virtuoso exercise of humane and common-sense reasoning brought to bear on house planning, studied in laboratory conditions. The central tool used throughout was the *plan*. »For it is the plan which in the simplest way defines the conception.« The central theme was a radical thinking through towards solutions.[1] Segal was impatient with housing researchers whose goal was not the search for optimal performance: »Optimal performance standards exist and should have been pursued with vigour.« The first major result of these laboratory exercises, *Home and Environment*, was a fascinating book of chess gambits, packed full of plans and layouts. As a quietly revolutionary document it is magnificent, and utterly down-to-earth, literally. It demonstrated conclusively, and in marked contrast to almost all its contemporaries, that humane low-rise and high-density residential areas are sensible and achievable goals. Segal played variations on this theme all his life; this book, with cool dryness, first stated his case.

Modernist housing

To make clearer the radical nature of his project – to centre housing ideas on »the small urban house« – a very brief résumé of twentieth century ideas in housing is worth repeating [I paraphrase it largely from an article by Segal.][2]

The accepted, and therefore usually unstated, intentions of Modern architecture were to improve the standard of the places we spend our time in, to improve the environmental conditions for everyone. Focusing down on the housing neighbourhood, then, one of the central themes was the achievement of a balanced condition, an equality of amenity standard (sunlight, privacy and so on) within the overall layout. Themes like this led to that archetypal »modern« housing solution the single row house. In Germany and Holland in the 1920s, the row house was much studied and designs refined, optimising orientation, sunlight and construction techniques. Double rows were developed, and then by the late 1920s tower blocks and slab bocks began to appear, at least as projects.

However the claims of the small house as a serious competitor in high-density housing were, as Segal points out, considered to be quite out of the question. The only real exception was in Holland where Oud's Kiefhoek terraces in Rotterdam demonstrated a real potential which remained a reference for many later architects.

By the early 1930s in Europe a variety in planning forms was already becoming possible, which led to the pattern known in post-war British housing as »mixed development«: here are large areas where, at least covertly for picturesque reasons, slabs and towers are interspersed with low rows of flats or houses.

An Amsterdam housing competition of 1933 was the first in which slabs and tower-block solutions were offered alongside double and single row terraces and low (4- to 6-storey) flats. One entry, considered important by Segal though completely overlooked both at the time and in the subsequent history of Modern housing, is H Leppla's proposal for two-storey houses at a density of 72 houses/acre (ca. 180 houses/ha). It was tight and tiny, but possible and humane. It was typical of Segal to have spotted and carefully studied it.

Thus the pattern which might have been expected after the war was »mixed development«, the design of housing areas which mixed tall flats with long slab blocks of flats and a residue of ground level housing for families. (Another strand, which Segal didn't even think worth discussion, was the highly serviced fortress, from

Ein Vorrat für den Wohnungsbau / Forschung für »Das Wohnen«

Im dunklen Jahrzehnt von 1938 bis 1948, als Walter Segal keine Gelegenheit hatte zu bauen, bestand sein wichtigster Beitrag zur Architektur in der »Forschung (...) und dem Studium von Entwurfsproblemen, ohne Begrenzungen durch die Forderungen eines Bauherrn.«

Segal beschrieb diese Arbeit als ein Auffüllen seines architektonischen Vorrats, seiner »Speisekammer«. Bald öffnete er diese dem Publikum, als er, von 1942 an, seine Studien zum Thema Wohnungsbau veröffentlichte, die dann 1948 im Buch *Home and Environment* erschienen. Diese virtuose Studie erörtert Argumente zur Planung von Wohnbauten gewissermaßen im »Labor«. Das zentrale Werkzeug sei der Grundriß. »Denn es ist der Grundriß, der am einfachsten die Vorstellung wiedergibt.« Segals zentrales Thema ist ein Durchdenken der Wege zur Lösung. Segal hielt nichts von den Wohnbau-Forschern, deren Ziel nicht eine optimale Erfüllung aller Anliegen ist. »Normen für solche optimalen Anforderungen gibt es, und man sollte mit aller Energie ihre Erfüllung anstreben.«

Segal spricht immer von »Planen«, nicht von »Entwerfen«, wie wir heute instinktiv sagen würden. »Planer« wird der Handelnde genannt, nicht »Architekt«, als ob selbst die Ausdrücke »Entwurf« oder »Architekt« zu weit von der Objektivität abweichen würden.

Home and Environment ist ein faszinierendes Buch, voll von Schachzügen, Grundrissen und Lageplänen, ein, in aller Stille, revolutionäres Dokument, gleichzeitig im wahrsten Sinne des Wortes fest auf dem Boden stehend. Es zeigte schlüssig, und im Gegensatz zu den meisten seiner Zeit, daß menschliche Wohnquartiere im verdichteten Flachbau ein vernünftiges und erfüllbares Anliegen sind.

Der Moderne Wohnbau

Um die Radikalität seines Projektes zu zeigen, lohnt es sich, einen ganz kurzen Abriß der Wohnbauideen des 20. Jahrhunderts zu geben, größtenteils aufgrund eines Vortrags, den Segal selber im Februar 1966 vor dem britischen Architektenverband hielt.

Das allgemeine Ziel der Modernen Architektur war es, den Standard der Orte zu verbessern, an denen wir den größten Teil unserer Zeit verbringen, die Umweltbedingungen für jedermann. Auf das Wohnquartier bezogen bedeutete das: Gleichwertige Bedingungen, gleiche Annehmlichkeiten, Sonne, Privatsphäre und so weiter für alle, innerhalb einer Gesamtanlage. Diese Themen führten zu einer archetypischen »modernen« Lösung, dem Reihenhaus. In Deutschland und Holland wurden in den Zwanziger Jahren diese Typen mehr und mehr entwickelt, die Grundrisse verfeinert, die Orientierung, die Besonnung, ebenso wie die Konstruktion optimiert. Es wurden Doppelreihen gebaut.

Dann, Ende der Zwanziger Jahre, erschienen nach und nach die ersten Turm- und Scheibenhäuser. Kleine Häuser jedoch, betonte Segal, als einer ernsthaften Konkurrenz im verdichteten Wohnbau, schienen nicht mehr in Frage zu kommen. Die einzige wirkliche Ausnahme waren die Kiefhoek-Reihenhäuser von Oud in Rotterdam, die noch einmal die wirklichen Möglichkeiten demonstrierten und zum Vorbild für manchen späteren Architekten wurden.

In den frühen Dreißiger Jahren waren in Europa schon eine ganze Reihe von Haustypen möglich. Das führte zu dem, was man dann im britischen Nachkriegswohnbau als »mixed development« bezeichnete: Es ging um große Flächen, auf denen, wenn auch nur um des »Malerischen« willen, einige Reihen niedriger Häuser zwischen Türme und Scheiben gestellt wurden.

In einem Wettbewerb in Amsterdam wurden 1933 zum ersten Mal Lösungen mit doppelten und einfachen Einfamilienhausreihen und vier- bis sechsgeschossigen Wohnbauten neben den üblichen Hochhausformen gezeigt. Eine Einsendung, die Segal besonders wichtig erschien, obwohl sie damals sowie auch später in der Geschichte des modernen Wohnungs-

Karl-Marx-Hof in Vienna to Quarry Hill in Leeds. One of the great post-war examples of this castle for the people was, of course, le Corbusier's in Marseille.)

The small house

Home and Environment centres on »the small house«. As a Continental European, Segal admitted being unprepared for the strength of feeling in England for the detached house and garden, and the residue of this astonishment never vanished. By the 1940s the place of the small house in urban areas needed redefining and adapting to contemporary demands of performance and efficiency. Here Segal saw a particular role for Britain in the post-war development of architectural ideas and he put the small urban house under the microscope with unprecedented clarity.

To Walter Segal, »the old truth still holds that nothing is so difficult to design as a good house.« *Home and Environment*, begins with precise, everyday realities listing the requirements of a home:

- two living rooms, if possible one facing the street and the other more private.
- direct access to back garden avoiding kitchen and/or living room becoming corridors and offering service entrance.
- refuse removal and service access easy from the street and not in back garden or alley.
- adequate storage for prams and bicycles which are bulky, need protection, and go out to the street.

Through designing, Segal studies
– *narrow frontage terraces* showing their pitfalls, designing an optimal 12ft (3.66m) frontage house with a central service band (stair and bathroom) top lit; and he shows how these houses, appearing very Oud-like could simply fit at 30 to the acre (or, 75 to the hectare).
– *three-storey terraces*, and he finds against them. (Elsewhere he argues, unfashionably as always, for three-storey flats rather than houses.)
– *terraced bungalows*, which are not so bad;
– *linked pairs of semi-detached houses*, which solve many of the problems of the other types;
– and, via what he calls the »compromise type«, he comes to the *linked L-houses*, either »castellated« with a patio on the street (not unlike what he built much later at Tasker Road, 1963) or with a private patio. (The train

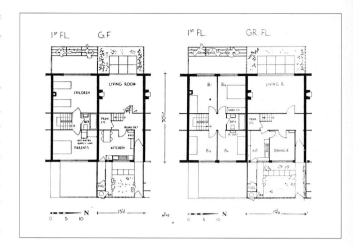

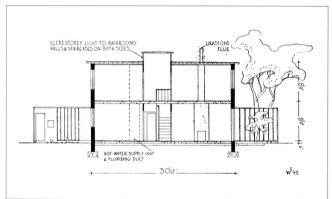

Design by Segal for 2-bed and 4-bed narrow frontage terraces, 1943. Plans, section, street view.
Entwurf von Segal für kleine Reihenhäuser mit 2 oder 4 Schlafzimmern, 1943. Grundrisse, Schnitt, Perspektive der Straßenfront.

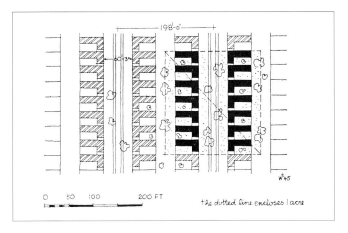

the dotted line encloses 1 acre

Design by Segal for L-shaped, linked single-storey houses, 1945. »14 to the acre«.
Entwurf von Segal für L-förmige, zusammengebaute, eingeschossige Häuser, 1945. 14 pro acre (ca. 280m² pro Haus).

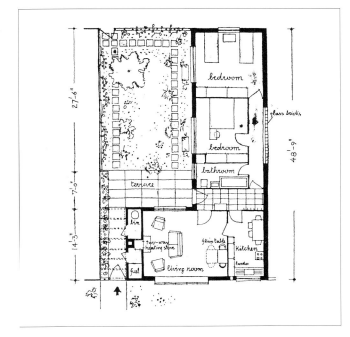

baus ganz übersehen wurde, ist H. Lepplas Vorschlag für zweigeschossige Häuser mit einer Dichte von ca. 180 Häusern/Hektar. Es war alles eng und klein, aber realistisch und menschlich vertretbar. Es war typisch für Segal, daß er diesen Vorschlag entdeckte und sorgfältig studierte.

Ein anderer Weg, den Segal nicht einmal einer Diskussion für würdig befand, waren die technisch gut ausgestatteten Wohnfestungen, vom Karl-Marx-Hof in Wien bis zum Quary Hill in Leeds. Das erste Beispiel dieser »Schlösser für das Volk« war die *Unité d'habitation* von Le Corbusier in Marseille.

Das kleine Haus

Home and Environment stellt das kleine Haus in den Mittelpunkt. Vom Kontinent stammend, gab Segal zu, war er gar nicht vorbereitet auf die englische Vorliebe für das freistehende Haus mit Garten. Dieses Erstaunen ist er niemals ganz losgeworden. Um 1940 war es an der Zeit, die Frage des kleinen Hauses im Stadtgebiet neu zu definieren und zeitgemäßen Forderungen anzupassen. Hier sah Segal eine spezielle Rolle für Großbritannien im Rahmen der Entwicklung architektonischer Ideen nach dem Krieg. Er betrachtete daraufhin das kleine städtische Haus gewissermaßen unter dem Mikroskop.

of thought which is best resolved in the patio house comes, in a sense, full circle stemming from the student competiton entry 15 years before.)

The linked patio house

To exemplify *Home and Environment*, the clearest essay in the study of house planning ideas until Chermayeff and Alexander's *Community and Privacy* of the 1960s, as well as to sample Segal's housing ideas in detail, I look briefly at this patio house (fig. p. 77 *left*).

Thirty, three-bedroom, L-shaped patio houses are formed round three sides of a residential block approximately double square in shape. To the street, there is a wall of housing set a short way back, articulated in a rhythm of two-storey blocks separated by slightly wider single-storey blocks with trellised terraces above. A generous path leads from the street to what is obviously each front door, a lesser one to the back doors in between.

Beyond the front door, which is overlooked from the dining table in the »active« living space, there is stair and toilet to the right, but the hall opens out and pauses under the stair before entering either of the two living spaces to the left.

The main living room, with large windows on two sides (to the south and to the east or west depending on location), is near enough square, with doors, windows and fireplace so positioned that it feels spacious, is not in any sense a corridor, and can feel the heart of the house. The fireplace is placed to encourage gregariousness rather than exclusion, but the whole room can easily be reoriented on a summer evening.

The dining room, leading from this, gives to the kitchen from which child's play can be supervised. In fact the kitchen is, as it were, the »bridge«. Reducing isolation for the cook, it is strategically placed to command views to the street and visitors (from over the sink), into the private patio and of the children (by looking over the worktop), to the dining room and to the back door.

Up the stairs, each bedroom has its particular qualities. The largest overlooks the back gardens (to east or west); the second opens onto its private terrace, giving to the patio which can be reached down a little garden stair; the smallest faces south into the patio.

The back entrance leads to a covered porch where children can play or washing be hung up in wet weather; there is space for rubbish, fuel storage, pram and bicycle or garden shed. Beyond is the »intimate gar-

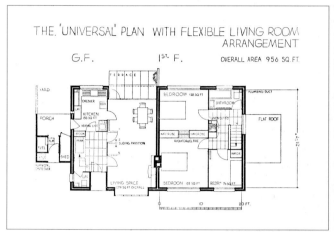

Design by Segal for linked semi-detached houses. The so-called »universal plan«, 1945.
Entwurf von Segal für »halbverbundene« Häuser. Sein sogenannter »Universalplan«, 1945.

den«, the patio, and to the right is the back door leading directly into the kitchen.

This subtle sequence of privacies inside is echoed in the sequence of external spaces. Both the back kitchen door and porch, and the diningroom's French window, give access to the patio; here the blank party wall (on the south or east side), along which the terrace steps come down, has a trellis to take creepers. This »intimate garden« is the house's outdoor room, and has an urban memory. Segal enjoyed quoting John Galsworthy's description of Soames Forsyte's London house as fitting this space of his: »Like the enlightened thousands of his class and generation in this great city of London... he inhabited a house which did what it could.... [At] the back (a great feature) a little court tiled with jade green tiles and surrounded by pink hydrangeas in peacock blue tubs. Here... the inhabitants or visitors could be screened from the eyes of the curious while they drank tea.«

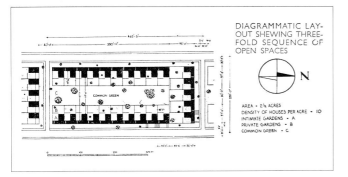

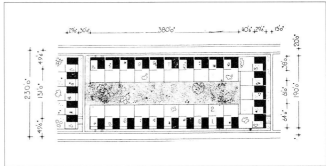

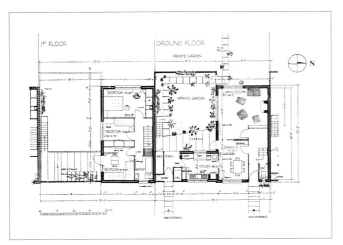

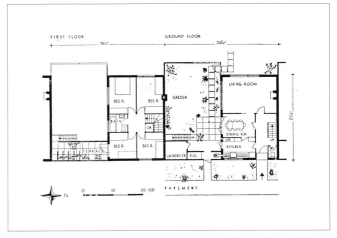

Segal's central theme is exemplified in this house type, the L-patio of 1945. Note the three-fold sequence of open space (middle).
Dieser Haustyp, das L-Patio-Haus veranschaulicht Segals Hauptthema (1945). Beachten Sie auf dem mittleren Bild die Abfolge der Aussenräume über drei Stufen.

Double rows of patio houses for any orientation, 1945. Plan is very similar to student competition project 1931.
Doppelreihen von Patio-Häusern für beliebige Orientierung, 1945. Dieser Grundriß gleicht sehr stark demjenigen, den Segal 1931 für den Studentenwettbewerb entworfen hat.

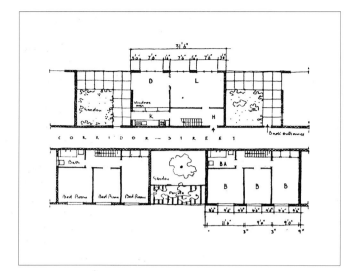

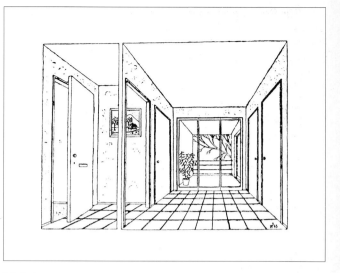

Design for elevation of maisonettes, 1942. Each dwelling with front doors off the brightly lit »paths« and back doors off the same path directly into their »hanging garden«. This contrasts with conventional, dark flat entrance (above).
Entwurf für gestapelte Maisonetten, 1942. Eine helle innere Straße erschließt die Wohnungen wie auch direkt »hängenden Garten« jeder Wohnung. Dies im Gegensatz zu den üblichen dunkeln Treppenhäusern und Gängen (oben).

Für Segal galt immer noch die alte Weisheit »daß nichts so schwierig ist, wie der Entwurf eines guten Hauses«. *Home and Environment* beginnt mit einer präzisen Liste der alltäglichen Anforderungen an ein Haus:

– Zwei Wohnräume, wenn möglich einer zur Straße und der andere eher ruhig gelegen.
– Direkter Zugang zum Garten, damit Küche und/ oder Wohnraum nicht zum Korridor werden, sowie ein Hintereingang.
– Müllabfuhr und Zugang für Dienstleistungen von der Straße, nicht von hinten über den Gartenweg.
– Ausreichend Abstellraum für Kinderwagen und Fahrräder, für die sperrigen Gegenstände, die Schutz brauchen, und mit Zugang zur Straße.

Schmale Reihenhäuser bieten dem Entwerfer Probleme. Optimal sind Häuser mit 12ft (ca. 4m) breiten Fronten und einem inneren, von oben belichteten Kern – Treppe und Bad. Von diesen Häusern, die von Oud sein könnten, gehen 30 auf einen acre (75 auf ein Hektar).

Dreistöckige Reihenhäuser lehnt er ab, obwohl sie in Mode seien.

Aufgereihte Bungalows sind nicht so schlecht.

Miteinander verbundene Doppelhäuser vereinfachen die Lösung mancher Probleme.

Gereihte Patio-Häuser

Als »Kompromißtyp« kommt Segal zu L- förmigen, aneinandergereihten Häusern, entweder »befestigt« durch eine Patio zur Straße (ähnlich dem, was er viel später, 1963, an der Tasker Road baute) oder mit einem Innenhof. Seine Überlegungen, die sich am besten im Patio-Haus verwirklichen lassen, führen ihn damit im Kreis zurück zu seinem 15 Jahre älteren Projekt im Studentenwettbewerb.

Um ein Beispiel zu zeigen aus *Home and Environment*, der klarsten Studie zum Wohnungsbau bis Chermayeffs und Alexanders *Community and Privacy*, greife ich diese Patio-Häuser heraus.

30 L-förmige Patio-Häuser, jedes mit drei Schlafzimmern, bilden drei Seiten eines Blocks, etwa doppelt so lang wie breit. Zur Straße hin bilden die etwas zurückversetzten Häuser eine »Mauer«, wechselweise aus zweistöckigen Blocks, getrennt durch etwas breitere eingeschossige Häuser, mit den durch Gitter geschützten Terrassen auf dem Dach. Ein breiter Weg

führt von der Straße zu den Haustüren, ein schmalerer zu den Hintertüren dazwischen.

Hinter der Haustür, vom Eßtisch im »aktiven« Wohnraum gut zu überblicken, liegen rechts die Treppe und die Toilette; die Halle öffnet sich noch einmal unter die Treppe, ehe man einen der beiden Räume links betritt. Der Haupt-Wohnraum, mit großen Fenstern an zwei Seiten, nach Süden und nach Osten oder Westen, je nach Lage, ist fast quadratisch; Türen, Fenster und Kamin sind so angeordnet, daß er geräumig, keineswegs als Korridor wirkt – das Herz des Hauses. Der Kamin lädt zur Geselligkeit ein. Der ganze Raum kann an einem Sommerabend auch andersherum orientiert werden.

Von diesem Raum aus kommt man ins Eßzimmer, dann in die Küche, von der aus man die spielenden Kinder überwachen kann. Die Küche ist gewissermaßen die Kommandobrücke. Der Koch ist nicht isoliert. Er ist strategisch plaziert, kann von der Spüle aus die Straße und etwaige Besucher beobachten, vom Arbeitstisch aus den Innenhof und die Kinder, dazu Eßzimmer und Hintertür.

Im Obergeschoß hat jedes Schlafzimmer seine eigenen speziellen Qualitäten. Das größte geht nach hinten, zum Garten, nach Osten oder Westen; das zweite öffnet sich auf eine private Terrasse, zur Patio hin, die man über eine kleine Treppe erreichen kann; das dritte ist nach Süden gerichtet, zur Patio.

Die Hintertür führt in eine gedeckte Veranda, wo die Kinder spielen und bei schlechtem Wetter die Wäsche aufgehängt werden kann. Dort ist Platz für alles Mögliche, Brennstoff, Kinderwagen, Fahrräder, Gartengeräte; anschließend folgt der intime Garten, die Patio. Die Tür rechts geht direkt in die Küche.

Die subtile Folge der Privaträume im Innern wird in den Außenräumen wieder aufgenommen. Über die Küchentür und die Veranda, oder durch die Fenstertür des Wohnraumes erreicht man die Patio. Die geschlossene Wand zum Nachbarn im Süden oder Osten, an der die Treppe von der Terrasse herunterführt, ist für ein Spalier mit Schlingpflanzen vorgesehen. Dieser intime Garten ist gewissermaßen das Zimmer im Freien, eine ausgesprochen urbane Form. Segal zitierte gerne John Galsworthys Beschreibung von Soames Forsytes Londoner Haus, das ihm auf diesen Raum zu passen schien: »Wie Tausende aufgeklärte Bürger seiner Klasse und Generation in dieser Großstadt London (...) bewohnte er ein Haus, das ihm bot, was nur möglich war. (...) An der

Through a »window« veiled by plants, this patio looks to the family garden, the next in the sequence of open spaces. That garden, overlooked by the living room, is reached by a gate in the corner of the patio. From here a path leads down the garden to the last in this sequence of spaces, a common green shared by these 25 dwellings, and enclosed by them to the north east and west, and overlooked across the minor access lane to the south by the kitchen fronts of the next block.

The levels of privacy are controlled with delicate balance: the common green clearly feels private to the block, but in fact it is part of the permeable structure of the neighbourhood and certainly not »secret«; the gardens of the houses seem exposed, but this is offset by every house also having an intimate patio, and anyway they could be planted as desired. Access to the patios from the open space – the way friends would come across the block, for example – forces a visitor to approach right up to the living room window before turning into the patio gate, which ensures exactly the appropriate level of privacy and deterrence.

There is no doubt that it is easy to inhabit these dwellings in the mind's eye simply by reading plans; to imagine real people, from young scampering children to elderly couples, enjoying filling them out as their habitations.

All this is in an essentially economic plan of a three-bedroom house of 943 sq.ft. (ca. 85m^2), with a living room of 330 sq.ft. (ca. 30m^2) – a considerably larger proportion of usable space than in a typical house of this size. Over two decades later, research at Cambridge by Leslie Martin and Lionel March conclusively showed such perimeter site planning, colonising the whole area in this way, to be also an inherently economical form of layout.

Other housing proposals

Segal then began to explore putting his houses on the ground in different ways, particularly colonising squarish plots *in* which the building sits, rather than traditional long thin terrace gardens. This wider theme developed towards the end of the book (called by Segal »layout technique«), begins to suggest a much looser, more even distribution of dwellings on the land. His »double-building line« layouts for semi-detached dwellings and four dwellings in a square, are among the most fascinating. Each house carefully colonises its open space, but keeps a distanced,

dynamic equilibrium between itself and the next building. This led directly to his experiments with even-tempered layouts, »carpet pattern« housing, in the sixties and seventies.

He argued (particularly in *Planning and Transport)* for towns to have high-speed traffic routes separate from large islands of inhabitation, divided by loose grids. Within residential areas, he proposed large building blocks, long rectangles up to 1000′ (305m) enclosing their own green »precincts«, and two or three building lines, set off from secondary access routes – and finally a pattern of small houses.

Though »the small house« always won the argument, Segal worked with an objective thoroughness to design convivial flats and maisonettes. With these, the layouts of individual dwellings was less problematic than the possibility of gentle, pleasant access. While *Homes for the People* made it very clear that internal access corridors were one of the conditions which would not be acceptable in post-war flats, this crucial point was wilfully overlooked by so many powerful architects during the next generation. Segal spent much time on halls and stairways and balcony access. Among the most interesting ideas are,

– first, a row of staggered flats, alternately handed, entered from a balcony. The flats – three or four storeys high – are therefore detached, almost like bungalows off a lane, lit from three sides and needing no windows facing the balcony.

– second, a similar idea for maisonettes, approached from a central corridor which, with the staggered units, is brightly lit and open. Next to each maisonette (with its own back gate from the »corridor-street«), is a double height garden (about 5.5m high), only screened from the lane by a lower wall. It is interesting to contrast this with the maisonettes of Le Corbusier of which that built for *l'Esprit Nouveau* 1925 pavilion is the most obvious. Segal stacks four units, Corbusier planned to stack three, and their way of dealing with facades is revealingly different. But the essential point is the different possibilities allowed by Segal taking the garden right back to the »street«(see figs. p. 78 *top*).

High-rise

By the mid-1950s, the inevitable trend was towards higher slabs and towers. Segal was aware that »above

COMPARISON OF SINGLE AND DOUBLE BUILDING LINES IN LAYOUTS OF EQUAL ACREAGES AND DENSITIES OF HOUSES

ASSUMED DENSITY = 12 HOUSES PER ACRE
OVERALL AREA PER HOUSE (two storeys) - 855 SQ.FT.

0   100   200 FT

W⁵ 1/45

DOUBLE BUILDING LINE (WIDE FRONT HOUSES)

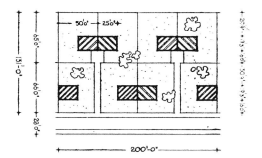

DOUBLE BUILDING LINE (MODIFIED TYPE; BLOCKS OF 4 HOUSES)

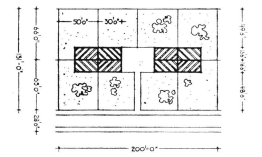

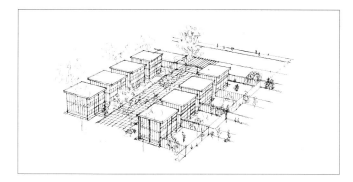

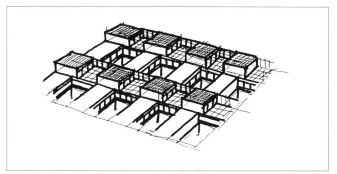

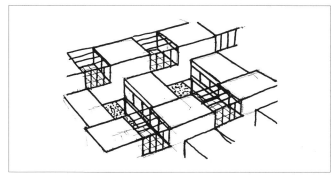

Left: »Double building line« layouts. These examples show semi-detached and blocks of four houses spread with a careful evenness over the neightbourhood. From »Home and Environment«, 1948.
Above: Later sketches for high-density/low-rise »carpet« houses. Examples of Segal's studies 1968-71. See also following page.

Links: Entwürfe für Doppelhaus-Reihen«. Diese Beispiele zeigen Doppelhäuser und Vier-Haus-Blöcke im Vergleich. Die unterschiedliche Anordnung verändert die Qualität der Außenräume. Aus »Home and Environment«, 1948.
Oben: Spätere Skizzen für hochverdichtete, niedrige Teppichsiedlungen. Beispiele aus Segals Untersuchungen 1968–71. Vgl. auch folgende Seite.

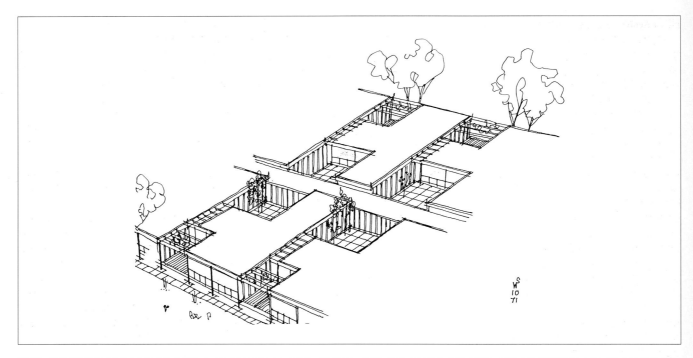

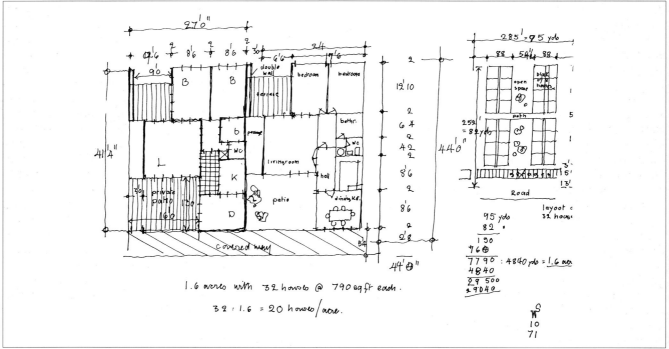

Further sketches for high-density houses from Segal's later studies.
Weitere Skizzen für verdichtete Teppichsiedlungen aus Segals späteren Untersuchungen.

Rückseite gab es einen kleinen Hof, der mit jadegrünen Fliesen ausgelegt und von rosa Hortensien in pfauenblauen Kübeln umgeben war. Hier (...) waren die Bewohner oder ihre Gäste vor den Augen der Neugierigen geschützt, wenn sie Tee tranken.«

Durch ein von Pflanzen verschleiertes Fenster blickt man von der Patio in den nächsten in der Abfolge von Freiräumen, in den Familiengarten. Diesen Garten, in den man vom Wohnraum hinaussieht, erreicht man durch eine Pforte in der Ecke der Patio. Von hier führt ein Weg den Garten hinab, bis zum letzten Raum dieser Sequenz, einer gemeinsamen Grünfläche für alle 30 Häuser. Er ist im Norden, Osten und Westen von diesen umschlossen; von Süden, über den schmalen Zugang hinweg, wird er von der Küchenseite der nächsten Hausreihe eingesehen.

Private und öffentliche Sphären werden hier in einem feinen Gleichgewicht gehalten. Das gemeinsame Grün empfindet man als private Sphäre des Blocks, obwohl es natürlich Teil der durchläßigen Struktur der ganzen Nachbarschaft und bestimmt nicht abgeschlossen ist. Die Gärten der Häuser erscheinen offen; aber das wird dadurch ausgeglichen, daß jedes Haus seinen intimen Innenhof hat; auch kann man den Garten nach Belieben bepflanzen. Der Weg vom gemeinsamen Grün zur Patio, der Weg also, den Freunde von der anderen Seite her kommen würden, zwingt den Besucher, bis zum Wohnzimmerfenster vorzudringen, ehe er durch die Pforte in die Patio treten kann. Das sichert genau den angemessenen Grad an Privatheit und an Abschreckung. Es ist einfach, sich vorzustellen, diese Häuser zu bewohnen, wenn man die Pläne liest. Man sieht wirkliche Menschen, von kleinen umherjagenden Kindern bis zu älteren Ehepaaren, Menschen, denen es Freude macht, diese Häuser zu bewohnen.

Der Grundriß für ein solches Haus mit drei Schlafzimmern und etwa 85m² Wohnfläche ist äußerst wirtschaftlich. Der Wohnraum von ca. 30m² ist im Verhältnis größer als üblich. Wie Leslie Martins Forscher in Cambridge mehr als zwei Jahrzehnte später schlüssig darlegten, ist auch die Umgebung wirtschaftlich geplant und angelegt.

Segal begann anschließend zu untersuchen, wie man die Häuser auch anders auf dem Boden verteilen könnte – vor allem bei der Besiedlung fast quadratischer Grundstücke anstelle der traditionellen langen, schmalen Reihenhausstreifen. Das umfassende Thema der Lageplantechnik, das er gegen Ende des Buches

in Angriff nahm, beginnt mit dem Vorschlag einer viel lockereren, gleichmäßigeren Verteilung der Häuser auf dem Grundstück. Seine Doppel-Baulinien-Pläne (die als Appendix zur zweiten Auflage, 1953, veröffentlicht wurden) für Doppelhäuser und für vier Häuser in einem Quadrat, gehören zum Interessantesten seiner Studien. Jedes Haus gestaltet sorgfältig seine Freiflächen, aber hält ein distanziertes dynamisches Gleichgewicht zum Nachbarn. Dieses Prinzip entwickelte er weiter bis zu den Teppichsiedlungen.

Obwohl das kleine Haus in der Bewertung am besten abschnitt, arbeitete Segal mit objektiver Gründlichkeit auch an Entwürfen für Geschoßwohnungen und Maisonettes. Hier war die Planung der einzelnen Wohnung weniger problematisch als die der Zugänge. Obwohl er in *Homes for the People* eindeutig klarstellte, daß der Zugang über Innenkorridore eines der Dinge ist, die für den Nachkriegswohnungsbau nicht mehr akzeptabel sein sollten, wurde dieser kritische Punkt durch so manchen »großen« Architekten der nächsten Generation übersehen. Segal dachte viel nach über Hallen, Treppenhäuser und Laubengänge. Zu den interessantesten Ideen gehören:

– Eine Reihe gegeneinander versetzter Wohnungen, abwechselnd nach der einen und der anderen Seite orientiert und von einem Laubengang aus zugänglich. Sie sind nicht aneinandergebaut, sondern fast wie Bungalows an einem Sträßchen von drei Seiten belichtet und haben keine Fenster auf der Zugangsseite.

– Eine ähnliche Idee für Maisonettes, die von einem zentralen Korridor aus erschlossen sind, der jedoch, wegen des Versetzens, hell belichtet und offen ist. Neben jedem Maisonette, mit einer eigenen Hintertür zur Korridor-Gasse, befindet sich ein Garten in doppelter Geschoßhöhe, etwa 6m, der nur durch eine niedrige Mauer gegen die Gasse geschlossen ist. Es ist interessant, diesen Vorschlag den Maisonettes von Le Corbusier gegenüberzustellen, zum Beispiel jenen, die Corbusier für den Pavillon *l'Esprit Nouveau* 1925 gebaut hatte. Segal stapelt vier Einheiten, Le Corbusier drei. Ihre Art, mit den Fassaden umzugehen, verrät, wie grundsätzlich verschieden sie waren. Das Wesentliche jedoch sind die Variationen, die Segal, indem er den Garten auch in diese Straßen brachte, anbot.

Hängende Gärten

Um die Mitte der Fünfziger Jahre ging der Trend unweigerlich in Richtung höherer Scheiben und Türme. Segal

five floors – and you can go up to 20 or 30 floors as is done nowadays – the social problems increase.« But again, while still arguing for dense separate houses, he attempted to improve on the high-rise he saw around. Access decks seemed rarely used and offered limited facilities for play. Nor did they much further neighbourly contacts, Segal felt. »So I tried to find a better answer.« In 1955, the new town of Hemel Hempstead commissioned Segal to design high-rise housing. His solution was to take two slim blocks and anchor between them large suspended platforms at various levels. In the end it was not built, nor again when a very similar scheme was proposed in 1964 for the LCC. Segal may, in retrospect, have been happy they remained on paper; but with their convenient planning, their host of little felicities in design, as well as the »hanging gardens«, they would surely have made more acceptable and remarkable dwellings than most British high-rise public housing (see figs. pp. 86/87).

The idea was to have 16 families around each »hanging garden«. »This could make a fairly workable community: the children could grow up together, and reasonable contacts would be established in a neighbourly way. In other words, you would take four units, four little houses, on either side of a suspended open space, arrange another eight houses on top, call the space between them a common green, and then repeat the operation every fourth floor.«

This gives an open space, 11m high, every fourth floor. The dwellings are maisonettes with plans reversed, which helps acoustic insulation and also means that every single dwelling was only seven steps up or down from the hanging garden. All dining-kitchens overlook the platform, as well as their private balcony which comes off the living room on the opposite face. For access there are fast lifts to each »hanging garden«. For escape the stair within the dwelling links the »back« door at bedroom level with the main hall, while beyond the house, the stairs linking to the »hanging garden« continue right down to the ground.

This 19-storey project was based on a regulation density of 136 people to the acre (330 to the hectare), and Segal felt sure that could be achieved with houses. He soon proved it with an unbuilt scheme for a London borough in 1964 of 3-storey narrow-front terrace houses at the same density, which was very much cheaper to build. So Segal began to take a wider look at the whole question of density and height of dwellings. Exercises in statistical geometry showed him how vertically stacked dwellings dramatically save land up to 3 or 4 stories; with gradually lesser savings until 8 to 12 stories, above which there is no argument at all on the basis of density.

So, for his lightweight housing method from the mid 1960s, Segal developed radically new »carpet« layouts. He proposed two-storey square houses, in a plot with private garden of similar size, grouped to give common open spaces and garages, at 22 dwellings and more to the acre or some 55 to the hectare (see fig. p. 81 *top*). »In most situations in the British Isles you will not have to exceed such densities. Therefore, I think that the detached house in a layout such as this can replace the terrace with its lower quality of amenity.«

But Segal's patrons had moved on; through the 1950s unassuming housing aims were quickly overwhelmed in the harsher, grander images housing architects and their clients developed. In the mid-1960s, as the force of low-rise high-density finally blew the cover of the high-rise fashion, an additional burden of up to 1.3 car-spaces per dwelling played havoc with Segal's tight patterns.

Looking back from 1969, Walter Segal said: »For architects in 1945 it became less and less possible to go on taking up where they had left off in 1939. Their visual world had been one of violence for too long to let them remain satisfied with the unobtrusiveness which the International Style had proclaimed.« British housing authorities needed the grand and obtrusive image. And when this mood was replaced in the 1970s it was not by modest and unobtrusive housing but by another image, from the saleable »hacienda« cluster to the Georgian »town-house«.

Notes and Quotations

1. Segal talks throughout of »planning« as the skill in question, rather than »design« as we might instinctively say today; »planner« (and the even more neutral »technician« used throughout *Homes for the People*) for the agent, rather than »architect«. As if even the terms »design« and »architect« stray too far from objectivity, from the study of housing centred on how it permits or inhibits our action.
2. Paraphrased largely from an article by Segal in *Arena*, March 1966, pp231–5.

Quotations from Walter Segal in this section from: 1945(B)3; 1960(B); *AJ* 2 March 1966; 1948(B); 1969(B).

war sich dessen bewußt, daß »oberhalb vom fünften Geschoß – und heute baut man bis zu 20 und 30 Geschosse – die sozialen Probleme zunehmen.« Obwohl er weiterhin für Einzelhäuser bei hoher Wohndichte plädierte, versuchte er gleichzeitig, den Hochhausbau, den er rings um sich sah, zu verbessern.

1955 erhielt Segal von der neuen Stadt *Hemel Hempstead* den Auftrag, Hochhaus-Wohnungen zu entwerfen. Seine Lösung war, zwischen zwei schmalen Blocks große Plattformen in verschiedenen Höhen zu verankern. Das Projekt wurde nicht gebaut, ebenso wenig wie ein sehr ähnlicher Vorschlag aus dem Jahre 1964. Rückblickend war Segal vielleicht ganz glücklich, daß alles auf dem Papier blieb. Dennoch: Dank der geschickten Planung, der Unmenge kleiner glücklicher Einfälle und dank den »Hängenden Gärten« wären es sehr annehmbare, sogar bemerkenswerte Wohnungen geworden, bessere als der Großteil des britischen sozialen Wohnbaues im Hochhaus.

Segals Idee war, 16 Familien um je einen »Hängenden Garten« zu gruppieren. »Das könnte eine einigermaßen funktionierende Gemeinschaft geben: Die Kinder könnten zusammen aufwachsen, und nachbarliche Kontakte würden entstehen. In andern Worten, man würde vier Einheiten, vier kleine Häuser an jeder Seite der offenen Fläche anordnen, acht weitere darüber, den Platz dazwischen als gemeinsames Grün bezeichnen, und dasselbe in jedem vierten Geschoß wiederholen.«

Das ergäbe einen freien Raum, ca. 12m hoch, in jedem vierten Geschoß. Die Wohnungen sind Maisonettes, die Geschoßgrundrisse wechselweise vertauscht – ein Vorteil für die akustische Isolierung einerseits; andererseits würde der Zugang zu jeder Wohnung damit nicht mehr als sieben Stufen, nach oben oder nach unten, vom »Hängenden Garten« entfernt sein. Alle Wohnküchen gehen zur Plattform, ebenso der private Balkon, der vom Wohnraum nach der andern Seite geht. Es gibt Schnell-Lifte zu jedem der »Hängenden Gärten«. Feuertreppen innerhalb der Wohnungen verbinden die Hintertüren vom Schlafgeschoß mit der Eingangshalle, während auf der andern Seite die Treppen zu den »Hängenden Gärten« sich bis auf den Boden fortsetzen.

Dieses 19-geschossige Projekt beruhte auf der Norm-Dichte von 136 Personen/acre (ca. 330/ha). Segal war aber sicher, daß man das mit Einzelhäusern auch erreichen könnte, und stellte es bald unter Beweis

mit einem ebenfalls nicht gebauten Projekt für einen Londoner Stadtteil, 1964: Mit dreistöckigen, schmalbrüstigen Reihenhäusern erreichte er dieselbe Dichte bei weit geringeren Baukosten.

So begann Segal die Frage der Wohndichte und Höhe der Gebäude in einem größeren Zusammenhang zu untersuchen. Übungen in Statistik und Geometrie zeigten ihm, wie man mit vertikal übereinandergesetzten Wohnungen mit bis zu drei oder vier Geschossen enorm viel Land sparen kann. Bis acht oder zwölf Stockwerke nimmt die Einsparung dann wieder ab. Darüber hinaus kann man auf der Basis der Wohndichte nicht mehr diskutieren.

So entwickelte Segal ab Mitte der Sechziger Jahre für seine Leichtbaumethode radikal neue Teppich-Pläne. Er schlug zweigeschossige quadratische Häuser mit gleich großen Privatgärten vor, so gruppiert, daß Raum für gemeinsame Freiflächen und Garagen bleibt, bei 22 und mehr Häusern/acre (55/ha). »In England sind größere Dichten in den meisten Fällen nicht notwendig. Deshalb glaube ich, daß das Einzelhaus in einer solchen Anordnung anstelle des Reihenhauses mit seinen erheblich geringeren Qualitäten und Annehmlichkeiten treten könnte.«

Aber Segals Auftraggeber gingen weiter. Bis in die Fünfziger Jahre gerieten die bescheidenen Wohnbauziele durch die großartigeren Vorstellungen der Wohnbauarchitekten und ihrer Kunden bald ins Hintertreffen. Und als Mitte der Sechziger Jahre die Überzeugungskraft des verdichteten Flachbaus endlich die Hochhausmode hinwegblies, kam als neue Bürde die Forderung nach bis zu 1,3 Autoplätzen pro Wohneinheit, eine für Segals dichte Teppichsiedlungen verheerende Forderung.

Zurückblickend sagte Segal 1969: »Für die Architekten wurde es ab 1945 weniger und weniger möglich, dort weiterzumachen, wo sie 1939 aufgehört hatten. Ihre sichtbare Welt war für zu lange Zeit von Gewalt beherrscht gewesen, als daß sie noch hätten zufrieden sein können mit der Unaufdringlichkeit, wie sie der Internationale Stil proklamiert hatte ...«

Die britischen Wohnbaubehörden brauchtes das großartige Leitbild. Und als diese Stimmung in den Siebziger Jahren endlich verflog, war es nicht, um zum Bescheidenen und Unaufdringlichen zurückzukehren, sondern es folgte ein anderes Ideal: die leicht verkäufliche Hazienda oder das Georgianische Stadthaus.

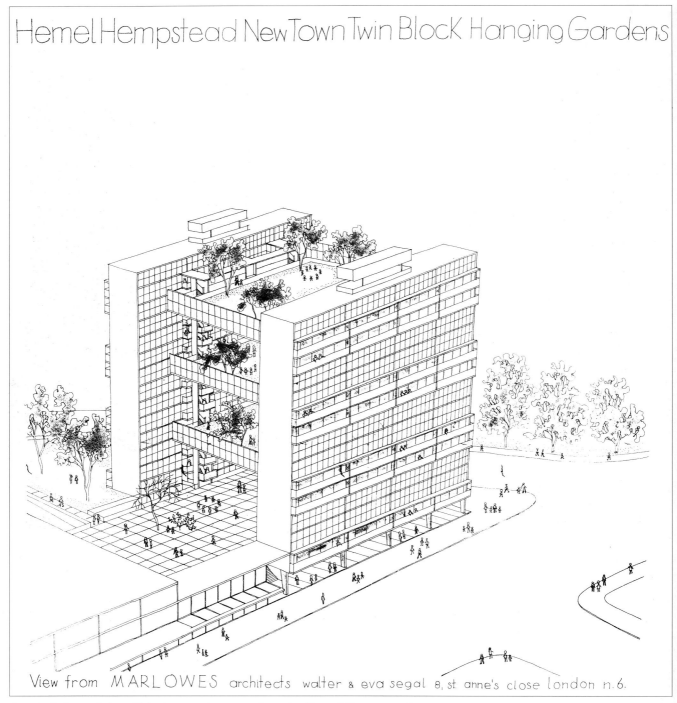

Hemel Hempstead New Town Twin Block Hanging Gardens

View from MARLOWES architects walter & eva segal 8, st anne's close london n.6.

General view of the »Hanging gardens« project for Hemel Hempstead, 1955.
Isometrie des Prokekts »Hängende Gärten« in Hemel Hempstead, 1955.

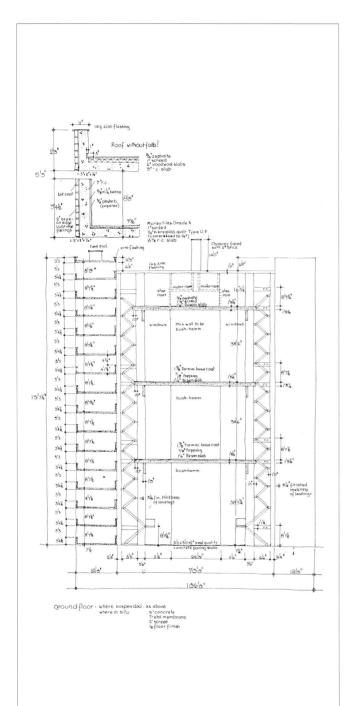

»Hanging Gardens« as developed for Lambeth, London, 1964. Axionometric from south-west (above) and section of block (left).

Projekt »Hängende Gärten« für Lambeth, London, 1964. Axionometrie von Südwesten (oben) und Schnitt durch den Block (links).

»Architecture«

We have seen Segal's understanding of the nature of building, and we have seen his deeply empathetic yet objectively investigative analyses of house form. But what else, beyond closely argued utlility and stability, did he mean by architecture?

The answer is not (to complete the Vitruvian triad of *commoditas*, *firmitas*, *venustas*) simply its visual attractiveness. For Segal, architecture must engage us at a deeper level than either simply visual charm or direct utility. There is, »quite separate from the evidence of efforts to please the eye, the essential meaning of shapes and arrangements which transcend utility and the enjoyment of the senses. This is perceived by the mind, which is wont to transfer it into a wider context for testing. Acceptance by the mind is essential for a building to be convincing.« Practical ability is not enough, and visual inventiveness is fleeting – it resists abstract rules such as proportioning systems. »To convince,« says Segal, »a shape needs more than proportions, texture or colour.« In the past, this element was supplied by symbolism, injected into that fusion of purpose and visual appeal to give it this deeper meaning. But today, without shared collective tenets, such symbolisms have very thin communicative value and are inappropriate.

Symbolism has been used through history, inevitably, to express power, with systems of axiality, directionality and clearly hierarchical architecture. Put at its crudest, »hierarchical space goes together with dictatorships, with kingships, with absolute rulers and so on. Non-hierarchical space is usually typical for freer societies. And that goes almost everywhere you have plans.

»The Roman town is based on a carfax, a crossing of *cardo* and *decumanus* in the forum. This space, where you see the axes which cross, surrounded by buildings, can be exceedingly depressing. Compare that with the Greek planned town; here also is an Agora (marketplace), but never street-crossing on an axis. Access to the agora is a long block lengthened; they would have space off the street, surrounded on three sides with collonade and public buildings, leaving the fourth side for traffic. So in the agora you could *meet*, people could come together. The Roman one was only for the Roman legion and administrators to march through. There are many contrasts like that – we come to the thinking of the Kriers in our own time!«

Segal had no sympathy for those, like his own former student Joseph Rykwert, who try to reawaken feelings for ancient rites. The historian's obsession with symbolic meaning, Segal felt, leads him, for example, to neglect the »regularly ordered but non-totalitarian and non-centralist Greek town« while praising the Roman one, so set about with ceremony. »To this much more desirable and human concept [i.e. the Greek city], which apparently needed no complex religious ritual, Professor Rykwert, alas, pays only scant attention.«

»Today, in our centrifugally explosive times, symbolic gestures no longer have the collective power or the permanence for them to add deeper meaning to our architecture.« About this, Segal has no nostalgia at all. »I think it stimulating that there is so much less common ground, and I see absolutely no reason why we should not accept this as a positive condition: a complex world generates complex environments.« And so, inevitably, the use of symbols today trivialises. »Indeed,« he adds, »there is nothing worse than shallow symbolism.«

While so many designers in the 1980s demonstrate an obligation either to ape symbolic gestures or else to compensate for the lack of them, an architecture which is »of itself« must aim to appreciate deeper-lying values. »This happens when a building can display a fuller understanding of its users than is achieved by mere serviceability, and when in turn it is likewise understood by them: in such reciprocal rapport a level of relationship is produced which is conducive to mutual well-being.«

And so an architecture which reaches the mind must have thought first for the body. Segal's personal phenomenology continues: »A handrail for a staircase may be made by a practical man and shaped by one that is visually conscious; but their combined effort does not make it more than a useful and good-looking article.

»Architektur«

Was versteht Segal unter »Architektur«?

Die Antwort liegt nicht (in Ergänzung zu Vitruvs *commoditas – firmitas – venustas*) einfach in der visuellen Attraktivität. Für Segal muß die Architektur anders ansprechen als durch visuellen Charme oder direkte Brauchbarkeit. Da gibt es den allen Formen und ihren Zuordnungen zugrundeliegenden Sinn, der weit über das Nützliche und über das Erfreuen der Sinne hinausgeht. Der Sinn wird durch den Geist wahrgenommen, der gewohnt ist, ihn in einen weiteren Kontext zu setzen. »Soll ein Gebäude überzeugen, so ist es unerläßlich, daß es durch den Geist, das Gemüt angenommen wird.«

Praktisches Wissen ist nicht genug, und visuelle Erfindungsgabe ist fließend – sie entzieht sich abstrakten Regeln. »Um zu überzeugen, braucht eine Form mehr als Proportionen, Textur und Farbe.«

In der Vergangenheit kamen solche Elemente aus der Symbolik. Sie vermittelte tieferen Sinn. Heute, da allgemeinverbindliche Grundbegriffe fehlen, hat individuelle wie kollektive Symbolik keinen verbindenden Wert und ist unzulänglich.

Symbolik ist in der Geschichte unverzichtbar gewesen, um Macht auszudrücken; mit Systemen der Axialität, der Ausrichtung und einer klar hierarchisch definierten Architektur. Um es ganz grob auszudrücken: »Hierarchische Räume gehören zur Diktatur, zum Königstum, absoluten Herrschern und so weiter. Nichthierarchische Räume sind in der Regel typisch für freiere Gesellschaften«, sagt Segal. (Tonband/84)

Der römische Stadtplan beruht auf dem *carfax*, einer Kreuzung von *cardo* und *decumanus* auf dem Forum, das heißt einem Platz, in dessen Mitte man die beiden sich kreuzenden Achsen wahrnimmt und der von Gebäuden umgeben ist. Das kann eindrücklich wie deprimierend wirken. Zieht man zum Vergleich eine griechische Stadt heran, so gibt es dort auch eine Agora, einen Marktplatz, jedoch niemals eine Straßenkreuzung in der Achse. Der Zugang liegt immer seitlich, über die Länge eines Blocks, und es gibt Raum außerhalb der Straße, mit Säulengängen und öffentlichen Gebäuden; die vierte Seite bleibt frei für den Verkehr. So kann man sich in der Agora treffen; Menschen können zusammenkommen. Die römischen Städte hingegen waren für Aufmärsche konzipiert. Wir kommen nicht umhin, an die Kriers unserer Zeit zu denken.

Segal hat keine Sympathie für die Besessenheit der Historiker für symbolische Bedeutungen. Das führe dazu, daß »die regelmäßig geordnete, jedoch nicht totalitäre und nicht zentralistische griechische Stadt« zugunsten der römischen, bei der die Zeremonien eine große Rolle spielen, vernachläßigt werde. »Dieses so erstrebenswerte und menschliche Konzept, das offensichtlich keine komplexen religiösen Riten brauchte, dem zollt Professor Rykwert, unglücklicherweise, nur am Rande Aufmerksamkeit.« (Segals Besprechung von Rykwerts *The Idea of a Town* in *The Architectural Review*, August 1977, ist recht scharf.)

»Heutzutage, in unserer nach allen Richtungen auseinanderfließenden Zeit, haben symbolische Gesten nicht länger die kollektive Macht oder die Beständigkeit, um unserer Architektur einen tieferen Sinn zu geben.« Segal war in dieser Beziehung gar nicht nostalgisch: »Ich denke, um so etwas zu stimulieren, gibt es zu wenig gemeinsamen Boden, und ich sehe absolut keinen Grund, das nicht als eine positive Voraussetzung anzunehmen: eine komplexe Welt schafft komplexe Environments.« Heute führt die Verwendung von Symbolen unweigerlich in die Trivialität. »In der Tat«, sagt Segal, »gibt es nichts Schlimmeres als eine hohle Symbolik.«

Und doch fühlen sich so viele Entwerfer verpflichtet, entweder symbolische Gesten nachzuahmen oder anderweitig ihr Fehlen zu kompensieren.

Aber eine Architektur, die ganz eigenständig, ganz »sie selbst« ist, muß dahin zielen, tieferliegenden Werten zum Durchbruch zu verhelfen. »Das geschieht dann, wenn ein Gebäude seinen Bewohnern mehr vermittelt, als nur nützlich zu sein, und wenn es andrerseits auch gelingt, das Verständnis der Bewohner für das Gebäude zu wecken. Eine solche Wechselseitigkeit der

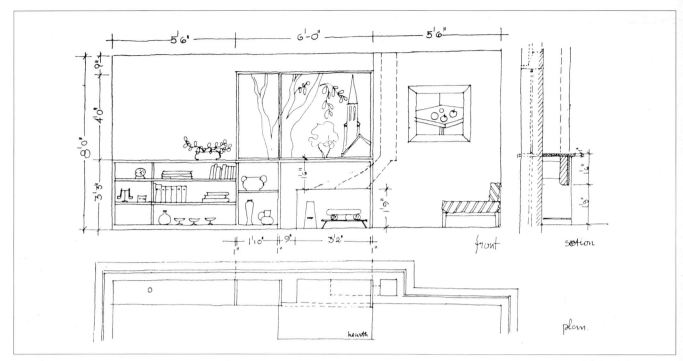

Sketch design for the end wall of the living room in Segal's own house, St.Anne's Close, London, 1950/52. Directly above the fireplace is a large window, as the flue bends to the right. This allows the living room to focus an the convivial fire but immediately undermines the weighty atavistic memories of a dominant hearth. Note the slender framing, the careful visual ordering, and the »Arthur Segal equi-valent fruit« on the wall (see p. 201).
Skizze für die Rückwand des Wohnraums in Segals eigenem Haus, St. Anne's Close, London, 1950/52. Direkt über der Feuerstelle befindet sich ein Fenster. Der Rauchabzug biegt nach rechts ab. So kann der Wohnraum auf das Feuer zentriert bleiben, ohne daß dieses den altertümlichen Charakter eines mächtigen Herds bekommt. Man beachte die Schlankheit der Konstruktion, die Sorgfalt der Gliederung sowie das Bild an der Wand: Arthur Segals »Früchte« von 1935 (vgl. S. 201).

If, however, there comes a third man, able to endow this handrail with a sympathetic response to the fingers that touch it, offering a welcoming and complementary shape to each, then the rail acquires a deeper meaning than mere practicality and visual concern can bestow on it.«

Segal, therefore, felt that the architect's duty is to study and understand how we interrelate in space and with the scenery with which designers form and fill space, »seeking knowledge about people's attitudes which transcend habit and condition, and often individuality as well.«

Not least in hinting at generalisability, thus at the possible construction of theory, this last point is important. Centrally, Segal sees part of the designer's role as to free us from »atavastic reactions.«[1] He talks of the design of places in which we are protected, can feel secure; in a word, places in which we »dwell«. Atavistic reactions of anxiety or agression can be called up, on the other hand, by simple thoughtlessness of design.

»The loss of a good place at the dining table will invite atavistic reactions, which often are irrepressible; exposure to excessive lighting may have similar effects; room colours may deprive a woman of her sense of security and stimulate negative impulses in her. Areas set aside in buildings for certain activities may lack differentiation and definition; often they are inadequately screened from circulation. In such cases a number of individuals are bound to be more privileged than others closer to the peripheries. Dissatisfaction will then assert itself, arising from the imbalance and inadequacy of conditions thus created by the designer's want of un-

Beziehungen kann zum allgemeinen Wohlbefinden beitragen.«

Aber auch eine Architektur, die den Geist erreichen will, muß zunächst an den Körper denken. Segal fährt fort: »Ein Handlauf einer Treppe kann von einem praktischen Mann gemacht und von einem andern formbewußt gestaltet worden sein, doch ihre vereinten Anstrengungen werden zu nicht mehr als zu einem nützlichen und gut aussehenden Gegenstand führen. Wenn jedoch ein dritter Mann hinzukommt, dem es gelingt, dem Handlauf zusätzlich etwas zu geben, das jedem, der ihn berührt, ein angenehmes sympathisches Gefühl vermittelt, ihn willkommen heißt, dann hat er dem Handlauf einen tieferen Sinn, über das Praktische und das gute Aussehen hinaus, gegeben.«

Segal glaubte deshalb, daß es Pflicht des Architekten sei, zu studieren und zu verstehen, wie wir in eine wechselseitige Beziehung zum Raum treten und wie der Entwerfer den Raum formt und füllt. Er muß versuchen, »sich Kenntnisse darüber zu verschaffen, wie Menschen sich zu dem stellen, was über Gewohnheit und Vorausgesetztes, ja oft über ihre Individualität hinausgeht.«

Walter Segal spricht über die Gestaltung von Orten, an denen wir uns geschützt und sicher fühlen, mit andern Worten, Plätze, wo wir verweilen mögen. »Der Verlust eines guten Platzes am Eßtisch kann atavistische Reaktionen auslösen, die oft nicht zu unterdrücken sind. Einem zu starken Licht ausgesetzt zu sein, kann zu dem gleichen Effekt führen. Die Farben in einem Raum können einer Frau das Gefühl der Sicherheit rauben und negative Impulse auslösen. Zonen in einem Gebäude, die für bestimmte Aktivitäten geplant sind, kann es an Differenziertheit oder Ausprägung fehlen; oft sind sie nicht genügend gegen den Durchgang abgeschirmt. In einem solchen Fall ist ein Teil der Anwesenden bevorzugt gegenüber denen, die sich weiter außen befinden. Unzufriedenheit entsteht dann aus dem Ungleichgewicht und der Unzulänglichkeit der Bedingungen, die durch das Unverständnis des Entwerfers entstanden sind, und Unkenntnis der Belastung, die er durch Nichtbeachten psychologischer Aspekte hervorgerufen hat. Geschützte Nischen vorzusehen, abseits vom Verkehrstrom, mit gleichen Annehmlichkeiten für alle, kann Bedingungen schaffen, unter denen atavistische Instinkte nicht geltend gemacht werden müssen. Es ist möglich, Leute um einen Kamin herum sitzen zu lassen, nicht nur um gleichermaßen im Warmen und gegen Zug und Vorübergehende geschützt zu sein, sondern auch, um alle an einer guten Unterhaltung teilhaben zu lassen, bei der jedem zugehört wird, wenn er etwas zu sagen hat, unbehindert durch einen größeren Nachbarn, der ihn verdeckt.«

Die Rolle des Entwerfers ist es deshalb, zu verstehen, wie er durch seinen Entwurf zu Geselligkeit, Ungezwungenheit und Angeregtheit beitragen und damit atavistischen Impulsen entgegenwirken kann.

Einer der wichtigsten Punkte, von Segal immer wieder betont, ist die »Chancengleichheit«. Sie ist ein Grundstein für gutes Wohnen. »Ein Gleichgewicht der Annehmlichkeiten für alle«, so drückte er es einmal aus. Das trifft nicht nur für eine Gruppe zu, die rings um den Kamin oder den Eßtisch sitzt, sondern mehr noch für die Lage der Häuser in einem Wohngebiet. Wie wir gesehen haben, geht es bei den Auseinandersetzungen und Gefechten in *Home and Environment* gerade um dieses. Hier wurden, ganz typisch, Lagepläne kritisiert, bei denen es um Eckhäuser mit zu wenig Abgeschlossenheit, um den Verkehr oder die Besonnung geht.

Was Segal mit Chancengleichheit meint, kann mit einigen weiteren Beispielen erläutert werden: Er sprach von Colette, die sich einen »sicheren« Stammplatz im Restaurant sicherte, von dem aus sie alles im Auge behalten konnte. Er vergleicht einen Tisch parallel zum Fenster mit einem im rechten Winkel dazu. Im ersten Fall »wenn die Aussicht gut ist, steht der Tisch falsch, und die Konsequenz ist oft, daß die Benachteiligten sich ganz bewußt irritiert fühlen, es sei denn, sie seien ihrer Umgebung gegenüber ganz unempfindlich, und das überträgt sich dann auf die Beziehung der Anwesenden zueinander und beeinträchtigt den Genuß des Zusammenseins. Dazu kommt noch etwas anderes: Die Gesichter der Leute, die mit dem Rücken zum Fenster sitzen, verschwinden fast vor der Blendung, während die der Gegenübersitzenden hell beleuchtet sind. Auf jeden Fall besteht ein Ungleichgewicht, das Machtverhältnisse begründen könnte.«

»An einem runden Tisch wird das Gespräch nicht abbrechen. An einem langen Tisch, selbst wenn man eine Unterhaltung zwischen all den verschiedenen Leuten erhofft hat, wird das Gespräch immer in zwei Hälften zerfallen, und die Leute werden mit der anderen Hälfte nicht in Kontakt treten ...« (Tonband/84). Und Segal kam wieder darauf zu sprechen, wie er an Bruno

derstanding and his unawareness of the strains produced by inattention to the psychological issues involved.

»The setting aside of sheltered bays, free from the stream of circulation, with a balance of amenity for all occupants, can ensure conditions in which atavistic instincts need not assert themselves.

»It is possible to make a fireside and seat people around it; not only to let each participate equally in the warmth, or to provide each of them with the same amount of protection from circulation and draught, but also to draw all of them into good conversation, where each will be listened to, if he has something to say, unhampered by the back of a bigger neighbour that hides him from view.«

And the role of designers, therefore, is to understand how design can encourage this gregariousness, ease and stimulation, and thus can successfully counteract atavistic impulses.

One of the obvious points from Segal's examples, clear in the words he uses too, is that equality of opportunity is a cornerstone of the creation of acceptable dwelling places. »A balance of amenity for all«. This is true not just of a group sitting round a fireplace or table, not just with a more complex interior spacial arrangement, but also in the layout of houses in a residential area. As we have seen, much of the thrust and argument in *Home and Environment* is precisely to that end. Typically he criticised layouts which included corner dwellings with poor privacy, circulation or sunlight.

The research Segal expected architects to undertake into how we dwell can be glimpsed in some of his numerous examples. He contrasts a dining table parallel with a window wall to one at right-angles to it. In the former, »if the view is good, the table is wrongly placed and the likely consequences are often consciously felt irritations on the part of the deprived – unless these are environmentally insensitive. This will, then, come out in the contacts between the two halves of the occupiers of the seats and interfere with mutual enjoyment.« There is also, of course, the converse point that those facing into the room have their faces somewhat hidden by the glare while those opposite them are brightly exposed. Either way there is an inequality, a power relationship potentially set up (see fig. p. 94 *top*).

»On a round table the conversation will be going! On a long table, even if you hope for conversation between all different people, you will always find the con-

versation dividing itself into two halves, and people do not make contact with those sitting elsewhere.« (Segal recalled sitting round Bruno Taut's round table, and Oedenkoven's in Ascona).

»Every individual has the desire to feel that space is there to express, and to be used by him or her, and that space must not overwhelm them. For example when they sit down, most people seek an area that is sheltered at the back. You can see exactly that with the behaviour of ordinary house animals. If a dog's bowl is placed in a circulation area, many an animal will delay starting his meal until this area is free of people; or else, it will eat looking anxiously behind between mouthfulls. A courageous dog, however, will reject having to eat without security, as he desires it, and retreat with his food to a sheltered corner, thus teaching his master where to place the bowl. Likewise, designers can be taught.« (See fig. p. 93.)

Visual scale and proportion

People are notoriously bad at judging the size of spaces with precision, Segal argued. Thus in his small buildings he indulged in an array of visual tricks to make spaces appear larger, by reducing door sizes and stair widths, or distorting other visual clues. He also was scathing of abstract visual rules, arguing: »design instinctively, visually, trust your eyes. And also let your clients, in the same way, decide.

»I learned fairly early to draw something that had agreeable proportions. In the days when I designed facades, I had an *image*. I drew facades, proportioned them with windows and whatever else, and then I began to measure *afterwards*. Which meant you could often produce highly romantic facades which were not regulated by geometry: the eye of a designer is *not* trained to follow; it is instinctive. So, when I'd drawn quite a number of facades I pinned them up on the wall and left a night in-between. In the morning, I came down and looked at them, and the one which took my eye, I decided to take and turn into a proper scheme. It was the first instinctive visual appeal. I assure you that I *never* chose the designs with ›exact‹ proportions – never!

»There is one other matter too: we are all born with different talents and abilities. Many people are highly musical, many people are visually sensitive; but the number of visually sensitive architects is very limited. You will find visual sensibility can, however, be

behind the door the assassin

»Few people like positions in which they can be surprised from behind.«
»Nur wenige Menschen mögen Plätze, wo sie von hinten überrascht werden können.«

there are deep shadows under her eyes

»Light can be a blessing or a curse; it should therefore be used knowingly.«
»Licht kann Freud oder Leid sein; es soll darum mit Kenntnis eingesetzt werden.«

a deep chair can be treacherous

»Sometimes the arms of a chair are preferable to the seat to secure the advantage of a good start.«
»Manchmal verschaffen die Armlehnen eines Stuhls den Vorteil eines guten und raschen Starts.«

the bowl in the wrong place

»Animals want to feel safe when eating.«
»Tiere wollen sich beim Essen sicher fühlen.«

(Four sketches and captions by Walter Segal, 1971 –
Vier Skizzen und Legenden von Walter Segal, 1971)

developed. But one thing can squash your talents and their development: and that is the whole problem of taste. Taste is a collectivist attitude which deprives you very often of your own more developed and talented ability to see things.

»I do think it is necessary not to let yourself be cheated out of your visual sensitivity, but to develop it, to see *more*; and hold to your own reactions to things.«

Segal would agree with C G Jung who, objecting to being given the role of guru, said: »I don't want to seduce anyone into believing, and thus take his experience from him.«[2]

Kitchen in terrace houses.
Küche in Reihenhäuser.

Notes and Quotations

1. Segal's position in the search for understanding behaviour in space is based not in reading, for example in phenomenology for which he had little time, but in his experience, his »strong animal confidence« (as he called it), and his acute observation.
2. C G Jung, letter to Prof. B Baur-Celio; 30 January 1934.

Quotations from Walter Segal in this section are almost all from 1973(B)3; also quoted are Wills & Yeo interview 1984, and book review in *The Architectural Review* August 1977.

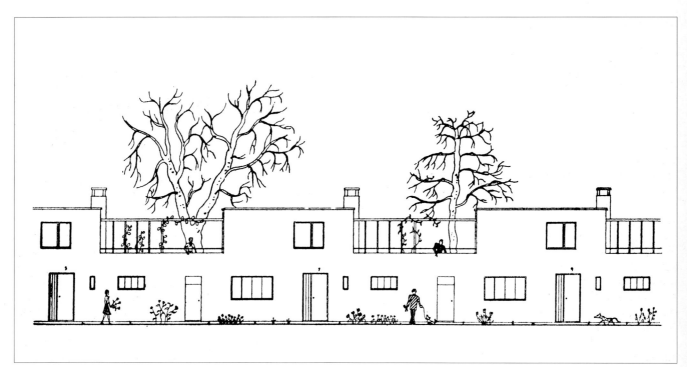

Dynamic equilibrium. Street elevation of linked patio houses, 1945.
Dynamisches Gleichgewicht. Straßenfront zusammengebauter Patiohäuser, 1945.

Tauts rundem Tisch gesessen hatte, und bei Oedenkovens in Ascona.

»Jedes Individuum möchte fühlen, daß Raum da ist, sich auszudrücken, ihn zu nutzen, aber nicht, um zu überwältigen. Zum Beispiel suchen viele Leute, wenn sie sich setzen, einen Platz, wo sie von hinten geschützt sind. Man sieht genau das auch am Verhalten gewöhnlicher Haustiere. Wenn der Freßnapf eines Hundes direkt in einem Durchgang steht, wird manches Tier warten mit dem Fressen, bis niemand mehr da ist, oder aber sich nach jedem Bissen ängstlich umsehen. Ein mutiger Hund jedoch wird es ablehnen, ohne Sicherheit essen zu müssen, und sich mit seinem Futter in eine geschützte Ecke zurückziehen, um so seinen Herren zu lehren, wo er den Napf hinzustellen hat. Auch Entwerfer sind lernfähig.«

Visueller Maßstab und Proportionen

Menschen können Maße im allgemeinen nur schlecht abschätzen, argumentierte Segal. Deshalb schwelgte er bei seinen kleinen Gebäuden in visuellen Tricks, um die Räume größer erscheinen zu lassen; er reduzierte die Türmaße, die Breite der Treppen oder verzerrte andere visuelle Elemente. Er spottete auch über abstrakte visuelle Regeln, mahnte: »Entwerft instinktiv, nach dem Aussehen, traut Euren Augen. Und laßt auch Eure Bauherrn in derselben Weise entscheiden.«

»Ich lernte ziemlich früh, etwas in angenehmen Proportionen zu zeichnen. In den Tagen, als ich Fassaden entwarf, hatte ich zunächst eine Vorstellung. Ich zeichnete die Fassade, gliederte sie mit Fenstern und was es sonst noch gab und fing erst hinterher an zu messen. Das heißt, ich konnte äußerst romantische Fassaden zeichnen, die nicht durch die Geometrie vorbestimmt waren: Das Auge des Entwerfers ist nicht ausgebildet, nachzuvollziehen; es handelt instinktiv. Wenn ich also eine Reihe von Fassaden gezeichnet hatte, steckte ich sie an die Wand und überschlief es eine Nacht. Am Morgen kam ich herunter, sah sie an und entschloß mich, diejenige, die mich auf den ersten Blick überzeugte, weiter auszuarbeiten. Es war der erste instinktive visuelle Eindruck. Und ich versichere Euch, daß ich niemals eine Skizze mit exakten Proportionen wählte, niemals.«

»Es gibt da noch etwas anderes: wir sind alle mit verschiedenen Talenten und Fähigkeiten geboren. Manche Menschen sind hochmusikalisch, andere visuell sensibel; aber die Zahl von visuell begabten Ar-

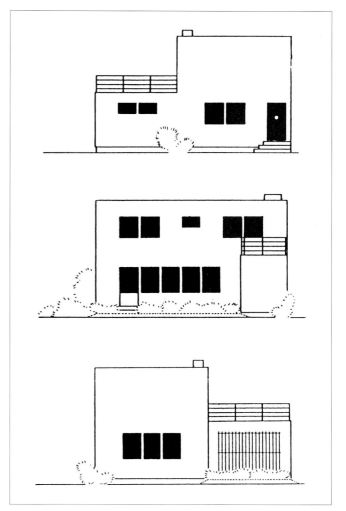

Student »small house« competition entry, 1931.
»Kleines Haus«, Studentenwettbewerb 1931.

chitekten ist sehr beschränkt. Ihr werdet jedoch finden, daß visuelle Sensibilität entwickelt werden kann. Aber etwas kann Eure Talente und ihre Entwicklung zerschmettern: und das ist das ganze Problem des Geschmackes. Geschmack ist eine kollektivistische Attitüde, die Euch oft um Eure eigene, besser entwickelte Fähigkeit, die Dinge zu sehen, bringt.

Ich glaube es ist notwendig, sich nicht um seine visuelle Sensibilität bringen zu lassen, sondern sie zu entwickeln, mehr zu sehen, und sich an die eigenen Reaktionen gegenüber den Dingen zu halten.«

The Post-war English Architect

As the war ended, the immediate prospects for architecture in Britain were no brighter than in the last years pre-war. Walter Segal gained relief from the tedium of a practice which had moved from air-raid shelters to bomb-damage surveys, by using teaching and writing as the vehicles for thinking.

He was writing articles on house design regularly from 1942, became a studio master at the Architectural Association in 1944, published *Planning and Transport* in 1945 and designed a didactic exhibition »Up your street«, specifically aimed to establish the link between the individual house and town planning, in the same year. The belief on which it was based, that post-war decisions would come from popular democratic processes, and that these decisions would be based on the rational good sense of an enlightened citizenry, was a general mood which Segal found particularly congenial. *Homes for the People*, his other work at exactly this moment, written as part of a left-wing group on behalf of their union the Association of Building Technicians, really is a »people's primer«, based on the premise that the educated client knows best. Despite his mass of well publicised housing ideas, Segal never built a housing authority scheme. A block of private flats in Lambeth begun in 1948 (see fig. p. 98 *bottom*) was followed by half a dozen small blocks of private flats in north and south-east London over the next decade. To the glance, these buildings appear very much of their moment: neat brick walls with large bands of glass under flat roofs. A second glance sees elevations enlivened by a careful geometric patterning of timber window framing; the closer look reveals a precision in the planning and in the appearance which commands attention.

The six-storey flats at Rutland Gate and Ovington Square, built in the mid-fifties on nearby sites in the centre of Victorian, fluffed-up, white stuccoed Knightsbridge, are economic and even-tempered.

Ovington Square (see fig. p. 97) shows Segal pushing material to its limit most elegantly. The stair, for example, is a 2 1/2′ (65mm) folded r.c. slab, and the balustrade of 1 3/8′ (35mm) diameter tubular uprights hold unframed trapezoid sheets of plate glass just at three points. On the exterior the window framing is, in effect, a complete framework in hardwood which is applied to the outside face of the brickwork. This ingenious and very simple layering produces an effect of depth on the facade as well as maximising depth and light in the rooms behind. When *The Architectural Review* called the front of Ovington Square »a textbook of bad manners«[1], Segal replied: »I am delighted to have out-

Flats in Rutland Gate, central London, 1954/56.
Kleiner Wohnblock Rutland Gate, London, 1954/56.

Ein englischer Architekt der Nachkriegszeit

Als der Krieg zu Ende ging, waren die unmittelbaren Aussichten für die Architektur in Großbritannien nicht besser als im letzten Vorkriegsjahr. Walter Segal war erleichtert, daß er nicht länger Luftschutzbunker zu bauen und Bombenschäden zu schätzen hatte.

Seit 1942 hatte er regelmäßig Artikel über das Planen von Wohnhäusern geschrieben; 1944 erhielt er einen Lehrauftrag bei der *Architectural Association*, 1945 veröffentlichte er *Planning and Transport* und entwarf die didaktische Ausstellung *Up your Street*, die

der nächsten Dekade, zuerst ein Block mit privaten Wohnungen in Lambeth (begonnen 1948), dann ein halbes Dutzend kleiner Mehrfamilienhäuser im Norden und Südosten Londons. Auf den ersten Blick wirken sie als Ausdruck ihrer Zeit: Saubere Backsteinwände mit großen Fensterbändern unter flachen Dächern. Aber diese Aufrisse sind belebt durch die sorgfältige Geometrie der hölzernen Fensterrahmen. In der Planung und im Aussehen kommt eine Präzision zum Ausdruck, die aufmerken läßt.

Flats in Ovington Square, central London, 1958.
Kleiner Wohnblock Ovington Square, London, 1958.

speziell darauf angelegt war, die Zusammenhänge zwischen dem einzelnen Haus und der Stadt darzustellen. Die Grundeinstellung, auf der sie aufbaute, nämlich daß die demokratischen Entscheidungen aus dem gesunden Menschenverstand der aufgeschlossenen Bürger wachsen würden, entsprach der allgemeinen Stimmung nach dem Krieg. *Homes for the People*, Segals anderes Werk in dieser Zeit, schrieb er als Beitrag für eine linke Gruppierung, die *Association of Building Technicians*; es ist eine Fibel für Laien, und beruht auf der Annahme, daß der gebildete Kunde weiß, was für ihn das Beste ist.

Obwohl er seine Ideen weiterhin veröffentlichen konnte, wurden in jener Zeit nirgends seine Vorschläge für Wohnsiedlungen verwirklicht. Bauten folgten erst in

Die sechsgeschossigen Wohnhäuser am *Rutland Gate* und *Ovington Square* aus der Mitte der Fünfziger Jahre liegen dicht beieinander im Zentrum des viktorianischen, aufgeplusterten, durch weißen Stuck geprägten Quartiers Knightsbridge und machen, im Gegensatz dazu, einen sparsamen und ausgeglichenen Eindruck.

Am Ovington Square sehen wir, wie elegant Segal die Materialien nutzte. Die Treppe, zum Beispiel, ist eine gefaltete, ca. 65mm dicke Stahlbetonplatte; und das Geländer aus Stahlrohrstützen, Durchmesser 35 mm, hält die ungerahmten trapezoiden Glasplatten an nur drei Punkten. Außen bilden die Fensterrahmen ein komplettes Hartholzraster über der Außenwand aus Backstein. Diese geniale und sehr einfache zweite Ebene gibt

raged you.« The visual geometry of this fascinating facade is most carefully patterned, the whole held together in an overall balance of rare precision. The image certainly belies the economical nature of the developer's brief.

Segal also built blocks in a suburban context which no more responded to the lower-density, Edwardian red-brick suburb than had the others to their tall stucco squares. The little block – virtually a cube – of four maisonettes at Compayne Gardens (1960) shows this at its best. The elevational geometry is direct but measured; again with struts proud of the walls to prevent the long sills from sagging – and so, of course, to keep timber sections to a minimum (see fig. p. 67).

The upper maisonettes have Segal's preferred solution of double-aspect double living space, with a terrace over the back gardens (which lights the stairwell) and a kitchen terrace to the street. The lower dwellings, blocked to the street by garages at ground level, are recompensed by magnificent living spaces. Double height for half their areas, there is a vast glass wall to the garden which, in its careful subdivision does not overwhelm.

The planning, as ever, is as tightly organised as possible; entering to the centre of the plan wastes the minimum in circulation. Plan, facades and spacial arrangement are a coherent essay in quietly exciting refinement.

Commercial buildings

While almost all his work is housing, in the mid-1950s Segal designed two commercial buildings of interest: offices and factory for a pickle maker in Hackney (east London) and headquarters for Tretol, a building products manufacturer in Hendon (north London).

Both were extremely cheap. Segal had built before the war for the Premier Pickle Co., and now they returned to him. The new factory was a very direct response to tight site and tight budget. It sat at the corner of two streets, hemmed in on the other two sides by existing buildings. On the corner Segal built a two-storey block, its rhythm of lightweight glass and tile walling covering brick piers is held between brick bookends. In here are squeezed canteen and toilets below, offices and caretaker's flat above. Behind this, where the pickles were cooked, bottled and stored, Segal laid out a »carpet« of uncut woodwool slabs and

Flats at Leigham Court, south London, shortly before occupation, 1959. Similar drawing of 1945 (above). Wohnblock Leigham Court, Süd-London, kurz vor Bezug 1959. Aehnliche Zeichnung von 1945 (oben).

der Fassade einen Eindruck der Tiefe und den Räumen dahinter ein Maximum an Licht. Als die *Architectural Review* (Januar 1959) diese Front am Ovington Square als ein »Muster schlechter Manieren« bezeichnete, antwortete Segal: »Ich bin entzückt, daß ich Sie erbost habe.« Die visuelle Geometrie dieser faszinierenden Fassade ist außerordentlich sorgfältig gestaltet, das Ganze wird durch ungewöhnliche Präzision im Gleichgewicht gehalten, und zeigt nichts von den knappen Mitteln, mit denen dieser Auftrag auszuführen war.

Segal baute auch im vorstädtischen Kontext Wohnblocks, die sich nicht länger an die geringe Wohndichte der im späten 19. Jahrhundert erbauten Vorstädte anpaßten. Der kleine Block – praktisch ein Kubus – von vier Maisonettes an den *Compayne Gardens* (1960) zeigt das am besten. Die Geometrie der Fassaden ist klar, aber maßvoll; auch hier Stützen, vor die Wand gesetzt, um das Durchhängen der langen Fensterbänke zu verhindern, und um die Holzquerschnitte auf ein Minimum zu reduzieren. In den oberen Maisonettes finden wir Segals bevorzugte Lösung: nach zwei Seiten ausgerichtete doppelte Wohnräume, mit einer Terrasse nach hinten, über dem Garten, durch den das Treppenhaus belichtet wird; Küchenbalkone zur Straße. Die unteren Wohnungen, die im Erdgeschoß durch die Garagen gegen die Straße abgedeckt sind, haben zum Ausgleich einen großartigen Wohnraum, der, in doppelter Höhe, die halbe Fläche beansprucht. Die große Glaswand zum Garten wirkt wegen ihrer sorgfältigen Teilung nicht dominierend. Der Grundriß ist, wie immer, so knapp organisiert wie möglich, der Zugang zur Mitte beansprucht ein Minimum an Verkehrsflächen. Grundriß, Fassaden und die räumliche Anordnung bilden zusammen ein Essay ruhiger und gleichzeitig erregender Raffinesse.

Gewerbebauten

Obwohl ihn vor allem andern der Wohnbau interessierte, entwarf er Mitte der Fünfziger Jahre auch zwei interessante Geschäftsbauten: Büros und Fabrikationsräume für eine kleine Konservenfabrik in Hackney (Ost-London) und das Hauptquartier für *Tretol*, ein Baustoffunternehmen in Edgware (Nord-London). Beide waren ganz billig.

Segal hatte schon vor dem Krieg für die Konservenfirma *Premier Pickle Co.* gebaut. Nun waren sie wieder zu ihm gekommen. Die neue Fabrik war eine direkte Antwort auf das knappe Grundstück und das knappe Budget. Sie lag an einer Straßenecke und war an den beiden verbleibenden Seiten durch bestehende Gebäude eingeengt. An der Ecke baute Segal einen zweistöckigen Block; die leichte Glas- und Ziegelwand vor den Backsteinpfeilern, ist zwischen zwei »Buchstützen«, Endwände aus Backstein, gespannt. In dieses Gebäude preßte er unten die Kantine und die Toiletten, darüber Büros und die Wohnung des Hausmeisters. Dahinter, wo einmal die Konserven gekocht, verpackt und gelagert werden sollten, legte Segal einen »Teppich« aus unbeschnittenen Holzwollplatten, und stellte dann einen Raster aus minimal gerechneten »Monitor«-Bindern als Tragkonstruktion auf. Diese ließen durch breite Streifen Licht von Osten und Westen einfallen, was zu einem guten Arbeitsklima beitrug. Es war ein leichtes und leichtgewichtiges Gebäude, das im Jahre 1958 weniger als 22 Pfund Sterling pro m² Nutzfläche kostete.

Die Kriterien bei *Tretol* waren ähnlich, die Kosten pro m² identisch. Das Programm war einfach und klar: Neue Büros, schnell errichtet, ohne unnötige Kosten, aber ohne Qualitätseinbuße. Segals Lösung war ebenso klar: Eine lange Wand, beidseitig Büros und Laboratorien, eine L-förmige Abschlußwand am hinteren Ende, eine etwas kostspieligere Eingangsfront am vorderen Ende, der offenen Ecke des Grundstücks. Ein dreistöckiger Treppenturm, mit dem Büro des Managers ganz oben, darunter den Toiletten, ganz unten dem Empfang, bildete den Blickfang des Ganzen, an den man später Erweiterungen anschließen konnte. 1962 baute Segal weitere Flügel an und einen dritten Stock im Leichtbau. Ein geplanter schlanker zehnstöckiger Stahlturm mit einem Büro pro Geschoß fand die Gnade der Baubehörde nicht.

Segal berechnete das Backsteinmauerwerk für drei Geschosse, Pfeiler aus silbergrauem Industriebackstein, Achsabstand ca. 1,60m. Die Rahmen standen auf einer Betonplatte und trugen die Stahlbetondecken. Typisch für Walter Segal war, daß er die Backsteine noch einmal selber testete, um zu kontrollieren, ob die Angaben des Herstellers auch stimmten.

Ebenso wie bei den Wohnbauten und der Fabrik gibt es auch bei den Tretol-Fassaden ein Gleichgewicht. Segal betont nicht die Vertikalstruktur oder die Geschoßlinien, sondern suchte das Gleichgewicht in der Hauptfassade. Die Fensterkonstruktion liegt vor den Pfeilern, und die gesamte Frontansicht ist damit eine einzige große Fensterwand. Diese durchgehende

Factory for a pickle manufacturer in east London, 1958. Monitor lights.
Konservenfabrik in Ost-London, 1958. Oberlichter.

Main building of the pickle factory.
Hauptgebäude der Konservenfabrik.

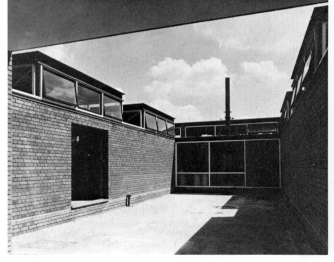

Internal court.
Innenhof.

(Fig. p. 101) / (Abb. S. 101) >
Pickle factory entrance, after many years use.
Eingang der Konservenfabrik, bereits viele Jahre in Gebrauch.

set up a grid of extremely minimal, calculated, »monitor« trusses across the site to support them. These let in bright strips of east and west light to give pleasant and unopressive working conditions; a light and light-weight building. All this was produced for under £ 22/m² in 1958.

The criteria at Tretol were similar; the cost/area identical – which (said *The Architects' Journal*) »would be regarded as low even in the housing field.«[2] In 1954, Tretol's needs were simple and explicit: new offices, quickly, without undue cost but without sacrifice of quality.

Segal's solution was straightforward. A long wall of double-banked offices and laboratory, an L-shaped bookend at the far end, a more expansive entrance front at the other, on the open corner of the site. This three-storey stair tower, with manager's office above rest rooms above reception space, was the focus

which would also hold together the projected extension.[3]

At Tretol as elsewhere, Segal stressed competence; in planning and in the speed and efficiency of its production. It was occupied within a year of the architect's commission. Segal credited this to: the client's prompt decisions when required; the local authority's helpfulness; working drawings being ready when the contract was signed; the architect having calculated all structural members; and the design being based on readily available material with minimal subcontracting.

He used calculated brickwork. Piers of silver-grey »semi-engineering« bricks, calculated to support three storeys, ran at 5 ft. (1.53m) centres along front and back with longer piers dotting a central spine. This frame stood on a concrete platform and supported concrete slabs.[4]

The Tretol building, front view (p. 102), and back view.
Tretol entrance hall.
**Das Gebäude der Firma Tretol, Vorderseite (S. 102),
Rückseite und Eingangshalle.**

Rahmenkonstruktion, fest in den Pfeilern verankert, ist sehr steif. Weil sie vor der Konstruktion hängt, bekommt das Holz genügend Luft und verrottet nicht. Von den Pfeilern ist nur wenig sichtbar; man sieht nur die Reihen der Kippfenster (nicht die Rahmen) und dazwischen die Reihen der 1,20x1,20m großen Ziegelplatten. Diese Ziegelplatten selbst waren ein Essay in ausgewogenem Design. »Aus dem Z kann man 22 Muster zusammensetzen, deshalb gewann es das Rennen. (...) Vom Spitzen des Bleistifts bis zum Abschluß der Montage brauchte man drei Monate.«

Das Dach war auch eine Pioniertat seines ungewöhnlich gewöhnlichen Verstandes: ein nasses Umkehrdach. Die Isolierung aus 50mm dicken Holzwollplatten, die man normalerweise als verlorene Schalung *unter* die Betonplatte legt, liegt *oben*, noch über den drei Lagen Dachpappe. Indem man das Dach ganz flach ausführte, mit Überlaufstutzen, die über die Dach-

fläche reichen, bleibt die Holzwolle bis zu einer Tiefe von ca. 19mm naß. Damit reduzieren sich thermische Bewegungen auf ein Minimum, kostspielige Klempnerarbeiten fallen weg, und der Bauvorgang ist sehr einfach. Einmal mehr, trotz des minimalen Budgets, gab es keinen Versuch – wie es damals unter Zeitgenossen üblich war – das härene Hemd brutal zur Schau zu stellen und auf jede Verfeinerung zu verzichten.

Moderne kleine Häuser

Aber im Zentrum von Segals Interessen stand weiterhin das kleine Haus. »Heute ein kleines Haus zu entwerfen, ist eine harte Aufgabe. Auch vor dem Krieg war es nicht gerade eine leichte, sie verlangte einen guten Teil an Begeisterung und – darf man hinzufügen – einen unendlichen Vorrat an Geduld. All das machte es zu einer Sache, der gut beschäftigte Architekten sorgfältig aus dem Weg gingen.«

Like the flats and the factory, Tretol's exterior has an »equi-valent« poise. Rather than express the vertical structure or the horizontal layering of floor slabs, Segal gives a »balance to the horizontal and vertical in the main elevation.« The window construction is super-imposed on the outer face of the piers, making the whole front elevation virtually one large window structure. This continuous framing, fixed securely to the piers, is very rigid. Hanging it all outside, ventilates the timber and minimises dry rot; minimal piers are exposed to the elements; and only rows of centre-hung pivoting windows are visible (no frames) with rows of 4 ft. by 4 ft. (1.22 × 1.22m) panels of tiles between (see figs. p.68 and p. 76 *bottom*).

The tiles themselves were yet another little essay in »equi-valent« designing: Segal played around to find a simple geometric tile pattern which could build into the most different designs. »The Z could make 22 patterns and so won the race.... From the first sharpening of the pencil to the fixed job on site, this was done in three months.« The roof, based of course on his uncommon common-sense, was a pioneering example of the »wet« and »inverted roof«. The insulation of 50mm woodwool slabs (departing from the normal practice of using them as permanent formwork under the r.c. cast slab) rests on three-ply felt on top of the slab. By casting the roof perfectly flat, and using an overflow upstand pipe, the wood-wool is kept wet to a depth of 19mm; the thermal movement is minimised, expensive guttering is omitted and drainage simplified. Once again, despite the minimal budget, there was no attempt (at that moment so popular among his contemporaries) brutally to display a hair shirt and to deny finesse; once again, this is exemplified in Segal's elegant stair design (see fig. p. 103 *right*).

Modern small houses

But the centre of Walter Segal's concerns was always in the small house. »Designing small houses nowadays is a tough proposition. It was not exactly an easy one before the war, requiring a good deal of enthusiasm and, may it be added, an infinite store of patience. All of which made it an affair which the busy practitioner was careful to avoid.«

He designed various individual little houses in the late 1940s, and, in this campaign to rescue the small house from the traditional local builder, in 1949 he edited a series of modern small house designs for *Good Housekeeping*.[5]

But, having been offered no projects to let him to put into practice his preferred housing solutions, with his wife Eva, Walter Segal decided to build for himself. They found a plot near where they lived, and with a teaching colleague, proposed a group of eight houses to be owned co-operatively by a group of friends. There were to be detached houses for the Segals and for their partner George Fairweather, an equally small, lively and intensely practical architect, a Scot who was senior teacher at the Architectural Association at the time. And there would be three pairs of linked semis in a row behind for the other six friends (see fig. p. 105).

The house planning came from *Home and Environment* and the layout was a »double-frontage« idea very much from that book. The »attached-semis« were entered as near the middle as possible to minimise circulation, each linked with porches which gave direct back door and garden access, and over which each stair projected so slightly. The wide-frontage living space was virtually glass walled, but so detailed as not to destroy its interiority or convey the feeling of living in a goldfish-bowl. While each had a fireplace and shelf unit carefully integrated into the end wall, in his own house Segal was amused to integrate a broad window most neatly above the fire. It was partly a little technical acrobatic; but it was also an attempt to wash away any atavistic infection which lingered round the tradition of »hearth« (see fig. p. 90).

Despite tightly screwed-down budgets, the buildings were carefully made, the »ordinary« appearance was controlled with carefully modulated and divided metal window frames and trellis-work, the facing bricks were handmade lavender-coloured bricks »like old Delft«[6]; the dark brown roof tiles were Belgian; the stairs of hardwood. Over all was a feeling of calm, modesty and gently shared open space.

But St Anne's Close, a central episode in Segal's life and career, also marked a period of prolonged trauma. Before work had got far, in the summer of 1950 Eva Segal, very pregnant and with failing eyesight and the fear that the sight might be lost altogether, tragically died. Walter Segal's world turned upside down. For three months he vanished completely. It was a year before he was seen again in the AA bar.

During his absence, Fairweather arranged the deeds for St Anne's Close into straightforward leases

St. Anne's Close. Segal's own house on extreme right.
St. Anne's Close. Segals eigenes Haus ganz rechts.

Während der späten Vierziger Jahre plante er verschiedene individuelle kleine Häuser, und führte einen Feldzug, um das kleine Haus dem traditionellen lokalen Baumeister zu entreißen. So verfaßte er 1949 eine Serie von Entwürfen von modernen kleinen Häusern für die Popularzeitschrift *Good Housekeeping* (Januar bis April 1949, Entwürfe von Walter Segal, Eva Segal, Neville Conder, Hugh Casson.) Nachdem ihm jedoch kein Auftrag angeboten wurde, eine seiner bevorzugten Lösungen in die Praxis umzusetzen, entschloß er sich, zusammen mit seiner Frau Eva, selbst zu bauen. Sie fanden ein Grundstück nicht weit von ihrer damaligen Wohnung und machten zusammen mit einem Kollegen den Vorschlag, acht Häuser zu bauen, die ge-

meinsam einer Gruppe von Freunden gehören sollten. Es sollten Einzelhäuser sein für die Segals und ihren Partner George Fairweather – auch ein kleiner, lebhafter und ungeheuer praktischer Architekt, Schotte und Lehrer an der Architectural Association. Dann sollte noch eine Reihe von Doppelhäusern für die andern sechs Freunde entstehen.

Die Pläne stammten aus *Home and Environment*, der Grundriß realisierte die »Doppel-Orientierung«, eine der Grundideen des Buchs. Die Eingänge zu den Doppelhäusern lagen so weit wie möglich in der Mitte, um die Verkehrsflächen auf ein Minimum zu beschränken, sie waren miteinander durch Veranden verbunden, von denen aus man direkt in den Garten gelangte, und

**View of one of the houses at West Heath Gardens.
Eines der Häuser in West Heath Gardens.**

**Inside the hall of a house at West Heath Gardens.
Eingangshalle eines der Häuser in West Heath
Gardens.**

which, while restricting alterations and including responsibility for common areas, entirely lost the social aims of a co-operative.

In the place of what may have started out with some memories of the qualities of communal life in Ascona, Segal returned to a life where his son had no mother, where he wanted to bring the boy up himself yet not become a full-time parent; where his partner Fairweather had moved out of the group to convert a large exisiting building nearby; and where he found himself, to his anger, landlord to the group which remained. Moreover he was architect to this group, now largely consisting of other architects and architectural critics; he was building their houses on extremely stringent budgets – the licenced size was 1000 sq.ft. (ca. 93m²), and the penal-ty for overspending the licenced money at that time (£ 22/m²)) was prison.

It was a recipe for trouble.

»The site resembled Passchendaele for most of the three years that I stayed there ... the contractors walking on or off the job.« said Alan Blanc, Segal's assistant after Eva's death.[7] The contract was a war of attrition. The first builder pulled out before losses bankrupted them, the second in litigation, and the third was not much less of a battle. Each week Segal saw the builder's acountant to agree prices and measure work done: »He made the meetings for Saturdays only, he turned off the central heating, opened the windows and sat down with the poor ac-countant and beat him down and down – on behalf

**Plan and street view of Segal's 1943 project for »castel-
lated« terrace houses.**
**Segals Projekt von 1943 für »aufgetürmte« Reihenhäu-
ser. Grundriß und Straßenfront.**

über denen jede Treppe leicht auskragte. Der große
Wohnraum war praktisch mit einer Glaswand abgeschlossen, die jedoch so detailliert war, daß die Ge-
schlossenheit des Innenraums nicht zerstört wurde
oder kein Gefühl aufkam, in einem Goldfischglas zu
sitzen. Kamin und Bücherregale waren sorgfältig in die
Endwand eingebaut. In Segals eigenem Haus gab es,
ein amüsantes Detail, ein Fenster gerade über dem
Feuer.

Trotz der knappen Budgets waren die Häuser sorgfältig ausgeführt. Ihr »gewöhnliches« Aussehen wirkte
diszipliniert und beherrscht durch sorgfältig gestaltete
und geteilte Fenster in Metallrahmen und Gitterwerk,
die äußere Backsteinschale aus handgefertigten, la-
vendelfarbenen Backsteinen, »wie alte Delfter« (Alan

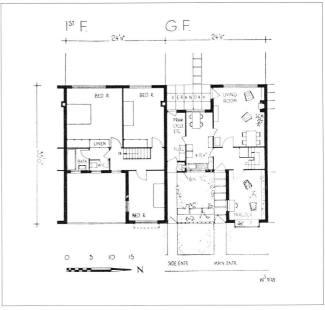

of all of us, of course!«[8] Some of the other residents, however, felt that Segal applied similar techniques with them; and when he submitted the final account with his usual precision but set out in neat spidery handwriting on a single print for each client from a large tracing, it was most unorthodox.

A decade after moving into St Anne's Close, Walter Segal and Moran Scott, who lived up the hill with her five young children in a tall, cold Victorian house, decided to marry. There was clearly no room in St Anne's Close for this new family. Segal designed an extension, a linked block of bed-sits for the children, and it was the last straw for the neighbours. In a vote, four of the eight vetoed the extension. Segal, astonished and hurt, had had enough and immediately began planning a new home elsewhere.

In the 1950s, Segal built a number of individual houses, notably a group of six at West Heath Gardens, Hendon, north London. They all followed his precept: »Simple shapes, few frills, a scientific use of traditional materials, cautious experiment with new materials. And yet higher standards of living and equipment in spite of austerity of architectural expression.« Leaving St. Anne's close in the early 1960s, Segal built his last north London brick houses: a terrace of three in Tasker Road, and his own in North Hill.

Tasker Road, which he and Moran developed speculatively provided three very compact and ingenious houses. The planning, while based on *Home and Environment*[9] is severely compromised by realities. The construction is fascinating and understated. Of the fully-glazed back wall one occupant remarked, »It was like living on a light shelf.«[10]

As always Segal is not out to astonish. *His* »less is more« is rather an overall economy, such as a network of minimum-section struts, rather than the »look no hands« gesture of a huge, pure span or opening.

Finally, his own house, »a fairly complicated house in traditional construction with a host of architects' details« (as he said), which he had begun planning in 1962, was complete in 1966. He had to produce four schemes before both they agreed (Moran vetoed a three-storey house) and the authorities approved. It took two months to demolish Moran's narrow, tall, Victorian interloper; it was replaced, in Segal's words, by »an open house on a much-walled-in site.«

It is one of his most ingenious and charming buildings.[11] The site planning, starting with the impossible 10′ frontage, is an essay in thresholds. The building is an amazing conjuring trick, getting a six-bedroom house, with two living spaces, separate garden access and an office, with a sense of spaciousness into the tiny area of a box 25′ by 27′ by 16′ (7.6 × 8.2 × 4.9m). Entered, of course, in the heart, the circulation space (Segal proudly calculated) is pared to 13% of the floor area (see fig. p. 112 *right*).

Careful sizing of openings accentuates scale. The house is full of unexpected, thoughtful details, from the most virtuosic suspended staircase (see fig. p. 146 *left*) to the simple double doors giving acoustic privacy between living and dining rooms (each stopping on expanded rubber joint filler dorstops). But in many ways it was the end of the line.

Notes and Quotations

1. January 1959.
2. *The Architects' Journal, 28 July 1955.*
3. In 1962, Segal added new wings as well as the originally planned light-weight third floor, and much later he designed a further extension. This tiny, tall, steel tower, of ten, one-room offices on top of each other, an economic, appropriate and pleasant working environment, was refused planning permission.
4. Typically, Walter Segal enjoyed testing the bricks himself to check the manufacturers' stated compressive strengths.
5. See *Good Housekeeping*, January to April 1949. There was a house by Walter Segal, one by Eva Segal (illustrated with a »WS« drawing) and others by Neville Conder and Hugh Casson. I am grateful to Neville Conder, who illustrated the series under the pseudonym Hugh Benson, for information.
6. Alan Blanc, »Working with Walter«, *Building Design*, 20 May 1988.
7. Ibid.
8. Monica Pidgeon, in conversation with the author, 27 February 1988.
9. Such as fig 47, p 33 (second edition). See figs. p. 107 this volume.
10. Peter Rich, »Notes on a Friendship« *A3 Times*, PNL, 1986
11. It is remarkable that it was never discussed in the architectural press.

Quotations from Walter Segal in this section from: *AJ* 15 January 1959, p.112-113; 1955(C)1; *Building* August 1949 p.284; 1953(B).

Blanc, *Working with Walter*, Building and Design, 20. 5. 88); die dunklen Dachziegel aus Belgien, die Treppen aus Hartholz. Über allem lag eine Stimmung von Ruhe, Bescheidenheit und zurückhaltender Gemeinsamkeit.

Aber St Annes' Close, wie die kleine Siedlung heißt, wurde in Segals Leben zum Trauma. Im Sommer 1950, bevor man mit der Arbeit weit gekommen war, starb Eva Segal auf tragische Weise. Sie war hochschwanger gewesen, gleichzeitig hatte ihr Augenlicht nachgelassen, sie drohte zu Erblinden. Walter Segals Welt brach zusammen. Für drei Monate war er verschwunden, und es verging ein Jahr, ehe man ihn wieder in der Architekten-Bar sah.

Während seiner Abwesenheit hatte Fairweather die Eigentumsverhältnisse geordnet, klare Baurechtsverträge aufgesetzt, die Änderungen einschränkten und die Verantwortung für die gemeinsamen Freiräume ordneten, jedoch die sozialen Ziele der Kooperative vollständig aufhoben.

Segal hatte das Unternehmen begonnen in der Erinnerung an die Qualitäten des gemeinschaftlichen Lebens in Ascona. Stattdessen kehrte er zurück mit einem Sohn, der keine Mutter hatte, mit dem Wunsch das Kind selbst aufzuziehen, ohne ihm seine ganze Zeit widmen zu müssen. Sein Partner Fairweather hatte die Gruppe verlassen und in der Nähe ein großes Haus gefunden, das er ausbaute, und Segal fand sich, zu seinem Ärger, zum Hausherrn für die restliche Gruppe ernannt. Die Gruppe bestand größtenteils auch aus Architekten oder Architekturkritikern. Segal hatte ihre Häuser mit äußerst beschränkten Mitteln zu bauen (£2/2/- per sq.ft.), bei einer Größe von 1000 sqft (ca. 93m²); die Strafe für Kostenüberschreitung war damals Gefängnis. Die idealen Voraussetzungen für Schwierigkeiten.

»Für den größten Teil der drei Jahre, die ich dort blieb, glich die Baustelle einem Kriegsschauplatz ... Die Baumeister kamen und gingen«, sagte Alan Blanc, der nach Evas Tod Segals Assistent geworden war. Der Bau wurde zum Zermürbungskrieg. Der erste Baumeister stieg aus, bevor die Verluste ihn in den Bankrott trieben, mit dem zweiten kam es zum Prozeß, und mit dem dritten ging es nicht viel besser. Jede Woche traf Segal den Buchhalter, bestimmte mit ihm die Preise und maß die ausgeführte Arbeit aus. »Er setzte die Treffen nur auf die Samstage an, stellte dann die Heizung ab und öffnete alle Fenster. So saß er da mit dem armen Buch-halter und drückte die Preise tiefer und tiefer – natürlich zum Wohl von uns allen.« (Monica Pidgeon im Gespräch mit dem Autor, 27. 2. 88)

Einige der Mitbewohner fanden jedoch, daß Segal ihnen gegenüber dieselben Methoden anwende. Als er die Abschlußrechnung mit seiner üblichen Genauigkeit präsentierte, aber in sauberer, dünner Handschrift, für jeden in einer einzigen Kopie als große Tabelle, wirkte das sehr unorthodox.

Ein Jahrzehnt, nachdem Walter Segal in St Annes' Close eingezogen war, beschlossen er und Moran Scott, die mit ihren fünf kleinen Kindern in einem großen, kalten viktorianischen Haus lebte, zu heiraten. Für diese neue Familie war in St Annes' Close natürlich nicht genügend Raum, und Segal plante einen Anbau mit fünf Wohn- und Schlafzimmern für die Kinder. Das brachte für die Nachbarn das Faß zum überlaufen. Mit vier zu acht stimmten sie gegen die Erweiterung. Segal, erstaunt und verletzt, hatte damit die Nase voll und begann sofort, ein neues Haus woanders zu planen.

Das eigene Haus

Das neue eigene Haus wurde »ein recht komplizierter Bau in traditioneller Bauweise, mit einem Haufen Architekten-Details«. Er begann es 1962 zu bauen, 1966 war es fertig. Er hatte vier Vorentwürfe gemacht, bis alle einverstanden waren (Moran lehnte ein dreigeschossiges Haus ab) und bis die Baugenehmigung vorlag. Es wurde, nach Segals Worten, »ein offenes Haus auf einem ummauerten Grundstück.«

Es ist eines seiner genialsten und schönsten Häuser. Situationsbedingt hat das Haus eine unmöglich geringe Frontbreite von wenig mehr als 3m. Es ist ein »Lehrstück der Übergänge«. Das Ganze gleicht einem verblüffenden Zaubertrick: Ein Haus mit sechs Schlafzimmern, zwei Wohnräumen, eigenem Gartenzugang und einem Büro in einer Schachtel von ca. 7,6 × 8,2 × 4,9m unterzubringen, und das mit einer gewissen Geräumigkeit. Der Eingang liegt natürlich in der Mitte, und die Verkehrsfläche beträgt, wie Segal stolz bemerkte, 13% der Geschoßfläche.

Sorgfältige Bemessung der Öffnungen betont die Maßstäblichkeit. Das Haus ist voll von unerwarteten, wohlüberlegten Details, von der virtuos aufgehängten Treppe bis zu den einfachen Doppeltüren zur akustischen Abschirmung zwischen Eß- und Wohnraum (mit einer Schwelle aus Schaumgummi-Fugenstreifen).

Tasker Road, North London. Street side (p. 110, above, right) with garages demanded by planning authorities .
Exit to gardens. Note the sophisticated engineering of the door: steel columns either side.
Häuser an der Tasker Road, Nord-London. Straßenseite (S. 110, rechts oben) mit von den Behörden vorgeschriebe-
nen Garagen. Zu beachten die raffinierte Konstruktion der Türe zum Garten: Stahlsäulen auf beiden Seiten.

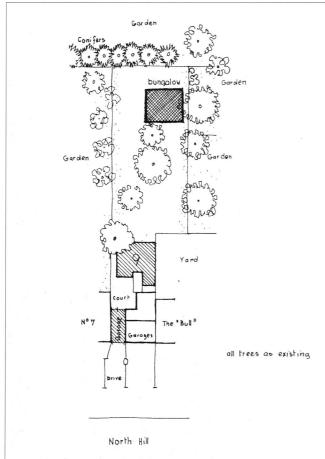

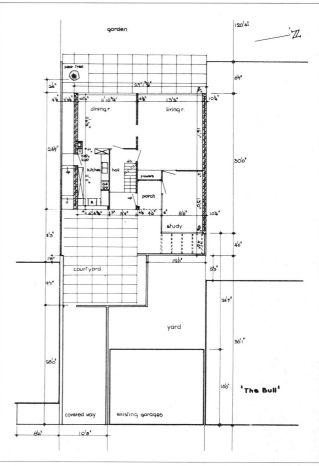

Segal's own house, Highgate, North London.
Location plan showing deep forecourt from the busy road (»drive«), the 3m frontage, and the »little house« named »bungalow« in the garden (see p. 130). Ground plan, showing the remarkably compact and yet open and unforced planning. This shows the subtle series of thresholds starting at the street. »I love my friends and the world, but I love my privacy,« Segal said. Yet his street door is not a tall defensive brick wall or the intimidating iron gates of some privacy lovers. It is just a low, bland, easily missed, garage front. There is a small side door but no bell to ring nor nameplate. The visitor must open this door, go through the covered storage room, then cross the open court to reach the porch. Here, overlooked by both kitchen and office, where the occupier is likely already to be waiting to greet you. If Segal wanted privacy, the street door could simply by bolted. Typically, it displayed an unagressive but firm and uncompromising understanding both of social boundaries and of physical site planning.

Segals eigenes Haus, Highgate, North London.
Die Situation zeigt eine Art Schutzschild gegen die stark befahrene Straße (»drive«), die nur 3m breite Vorderfront sowie das »Kleine Haus«, benannt »Bungalow« im Garten (vgl. S. 130ff.).
Der Grundriß zeigt eine Abfolge von subtilen »Schwellen«. Segal sagte: »Ich mag meine Freunde und die Welt, aber ich mag auch meine Privatheit.« Den Abschluß zur Straße bildet ein Garagentor. Daneben befindet sich eine schmale Seitentür, an der jedoch weder Klingel noch Namensschild auszumachen sind. Nur eine Briefklappe weist auf Bewohner hin. Der Besucher muß diese Tür öffnen und dann durch den gedeckten Gang gehen, worauf er in einen Hof gelangt. Dort kann er sowohl von der Küche wie vom Büro aus (vgl. S. 114) beobachtet werden und muß sich entschließen, ob er sich Richtung Haus oder Büro wenden will. In jedem Fall ist er nun berreits wahrgenommen worden und wird alsogleich begrüßt werden. Wenn Segal seine Ruhe haben wollte, verriegelte er einfach die Straßentür.

**Segal's own house. Entrance, living room and garden front with »Little House« (left), photographed 1980.
Segals eigenes Haus. Eingang, Wohnraum und Gartenseite mit »Kleinem Haus« (links), fotografiert 1980.**

Principles of Practice

Walter Segal began his architectural career working on one small project, on his own, producing the drawings very fast, freehand, in clear pencil line, doing the engineering calculations required and supervising the craftsmen's construction. Thus was the pattern set. As his career unfolded over the years, with an almost Zen ability, he managed yet further to minimise wasted energy. Aiming to balance his life, Segal strictly controlled the boundaries of his practice, banished wasted effort and unwelcome intrusion, leaving himself freedom to concentrate and to relax; to play.

When he was 45, Segal reached a crossroads in his practice: Was he to take the usual path of building up an organisation, a rolling work-load programme, rising overheads, and so on; increasingly divorced from the drawing board where, even were he able to hold onto standards, it could only be at an unacceptable cost to his whole life. Instead, turning down a large project he parted from his last assistant, deciding »to stay a small practitioner with a selective practice, to retain the joy of intensive designing and building with all it entails not omitting the variety of human contacts. For me there is more joy in seeing a smaller building through from beginning to end than playing a limited part in the construction of a large design.«[1]

»The most typical aspect of my practice,« he said in 1968, »is the firm desire to exercise complete one-man control. This involves the continued effort of extending such control over as many fields as possible by acquiring more and more knowledge and by taking more and more responsibility. With regard to the latter I should say here that I am one of the few practicing architects in this country who is not insured against claims for negligence.« With everything under

JUSTICE AND THE ARCHITECT

by Walter Segal; drawings by Kenneth Browne

Justice and the architect, drawing by Kenneth Browne. »Gerechtigkeit und Architekt«, Zeichnung von Kenneth Browne.

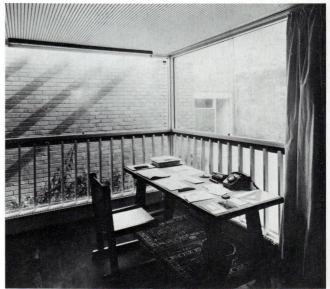

Segal's office at his Highgate house, London. Segals Büro in seinem Haus in Highgate, London.

Prinzipien des Berufslebens

Segal begann seine Laufbahn als Architekt mit einem kleinen Projekt. Er zeichnete alles sehr schnell, freihändig, mit klaren Bleistiftstrichen; er führte die verlangten statischen Berechnungen aus, beaufsichtigte als Bauführer die Handwerker. Damit war ein Muster gesetzt. Während seiner ganzen Karriere, über all die Jahre, lernte er mehr und mehr, mit einer gewissen Zen-Fähigkeit, jegliche Verschwendung von Energie zu vermeiden. Er brachte sein ganzes Leben in ein Gleichgewicht. Durch genaue Beachtung seiner Grenzen, Verbannen jeglicher vergeblicher Anstrengung und unwillkommener Störungen, bewahrte er sich die Freiheit, sich abwechselnd zu konzentrieren, zu entspannen, zu spielen.

Als er 45 war, hatte sich Segal zu entscheiden: Wollte er den üblichen Weg gehen, eine Organisation aufbauen, wachsende Arbeitslast, immer mehr Administration und so weiter auf sich nehmen? Immer weiter sich vom Zeichenbrett entfernen, sich verzetteln und immer weniger sich um die Details kümmern? Und selbst, wenn er seinen Standard aufrecht erhalten könnte, zu welchem Preis für sein ganzes Leben? Er entschied sich eindeutig, lehnte ein großes Projekt ab, trennte sich von seinem letzten Mitarbeiter und beschloß »ein kleines Büro mit einer selektiven Praxis zu bleiben, die Freude des intensiven Entwerfens und Bauens mit allen Folgen, nicht zuletzt der Vielfalt menschlicher Kontakte, sich zu erhalten. Mir macht es mehr Freude, ein kleines Gebäude von seinen Anfängen bis zum Ende zu begleiten, als nur einen kleinen Teil beim Bau eines großen Projektes zu bearbeiten.«

»Der typischste Aspekt meines Berufslebens ist« – sagte er 1968 – »die feste Absicht, alles ausschließlich unter meiner eigenen Kontrolle zu haben. Das bedeutet ein ständiges Bemühen, diese Kontrolle auf so viele Gebiete wie möglich auszudehnen, indem ich mir mehr und mehr Wissen aneigne und mehr und mehr Verantwortung übernehme. Deshalb bin ich auch einer der wenigen Architekten in diesem Lande, die nicht gegen Forderungen wegen Nachlässigkeit versichert sind.«

Segal war überzeugt, daß, wenn er selbst alles unter Kontrolle hätte, einfach keine Möglichkeit für Nachlässigkeiten bestände, und er betrachtete das als einen angemessenen Ausgangspunkt für seine Laufbahn. »Seinen Weg allein zu gehen«, sagte er ein andermal, »schärft den Instinkt.«

Segal hat nur kurze Zeit für andere gearbeitet, und das war für ihn fast so frustrierend wie die Zeit, als er selbst Hilfskräfte hatte. Anläßlich der Besprechung von E.F. Schumachers *Small is Beautiful* bemerkte er: »Für uns Architekten ist die These des Autors, daß die Kreativität in einem kleineren Bereich stärker angeregt wird, eine tägliche Erfahrung, die wir eigentlich offen zugeben könnten, da wir es ja alle wissen.«

Seine hart erkämpfte Freiheit brachte Segal mehr Zeit für das Entwerfen. Alle seine Rationalisierungen im Bürobetrieb waren »ausgerichtet auf die Notwendigkeit, mehr Zeit zu haben, Dinge zu durchdenken. In früheren Jahren konnte ich einem Kunden nicht sagen, daß ich noch nicht alle Möglichkeiten ausgeschöpft, noch keinen optimalen Plan hätte. Ich war gezwungen, zu früh Entscheidungen zu treffen, und das lasse ich jetzt nicht mehr geschehen.«

Der Entschluß, allein zu arbeiten, erlaubte es Segal, nach seinem natürlichen Rhythmus zu leben. Er war gerne produktiv, arbeitete ungestüm, doch mit bemerkenswerter Genauigkeit; aber er meinte, das könne man nicht nach der Uhr tun, weder er noch jemand anders. Während er so viel Zeit wie möglich für das Entwerfen freimachte, blieb bei diesem Vorgang doch etwas Geheimnisvolles. »Ich halte nicht an der ersten Idee fest, selbst wenn es eine Eingebung ist, überzeugend scheint. (...) Die Zeit, in der man die Dinge langsam brodeln läßt, ist eine sehr nützliche ...«

Wenn man alleine arbeitet, ändert das die Einstellung, wie beim Spiel der Kinder oder dem Handwerker in Mallorca, der einen Tag lang schwimmen ging oder einfach das Schmiedeeisengitter »nicht machen mag«. »Als ich noch einen Assistenten hatte«, sagte er zu mir 1976, »hatte ich immer etwas, das ich schnell über das Buch, das ich gerade las, schieben konnte, wenn er

this control there would simply be no chance of negligence.

»Going it alone,« he said another time, »sharpens the instincts.« Segal only briefly in his life assisted others and he felt nearly as frustrated when he had assistance himself. »For us architects the thesis that creativity is better stimulated within a smaller compass, is daily experience which might be more freely admitted since we all know about it.«

His hard-fought freedoms did bring Segal a greater share of time for designing. All his rationalisations of office practice were »conditioned by the need to have more time to think things out. In former years I could not tell a client that I had not yet exhausted the possibilities, that I didn't have an optimal plan. I was pressed for decisions too early; but now I don't let that happen.«

Having decided to work on his own allowed Segal to respond to his natural pattern of concentration. He enjoyed being productive, working furiously with remarkable thoroughness; but this could not follow a clock-bound routine, for him or (he felt) for anyone. While it freed up as much time as possible for design, that process itself nevertheless remained mysterious. »I do not settle on the first idea even if it's an insight, even if it seems convincing.... The period of letting things simmer is a very useful one.« Segal was very aware that clients, committees or public bodies especially, move at a certain pace and that the goal of minimising time, of honing down with systematic moves, cannot work if out of step with the client. Moreover, the brief for a building tends to grow gradually, often with addenda which become formative to the project. There is therefore no point forcing it.

Working alone allows changes in focus; exactly as in child's play, or consider the Mallorcan craftsmen, who spent a day swimming or »didn't want to« make Segal's wrought-iron gate. »When I had an assistant,« he admitted, »I always had something to slide over the book I was reading when he came in. Even today, I'm morally vulnerable if Moran catches me reading a book when I'm under pressure.... No-one ever told me that this routine was required of an architect.«

»With the degree of self-protection which I give myself I remain unharrased. There are other things in life besides being an architect. The one-man firm takes time to build up. It does not travel the road that leads to riches, or to the limelight. But it makes friends that return and, above all, it leaves scope for leisure and living.« He could not understand those who burned themselves out through overwork, dedication and seriousness. »Le Corbusier ran in the Bois de Boulogne at breakfast time when I'm in bed, he painted through the morning which I would have spent reading the papers, and he turned to architecture and his office in the afternoon.« Of the elderly Aalto, Segal asked, »Wouldn't it have been nicer for him if he had been able to go on producing fewer, but better buildings?«

»The most important problem for an architect is to know both his purpose and his limitations, his ambition and the range of his talent, and to balance his gifts and inadequacies. In my case I desire freedom in my work above all and I had to learn to accept only as much work as I could undertake without delegating responsibility.«

»How then does a one-man firm run?« The answer is: »by simplification, shortcuts, extreme centralisation and a sharp eye on economy. The one-man practitioner by running his contracts himself has much closer contacts with the building team, e.g. with quantity surveyors, engineers and technicians, specialists, clerks of work, contractors and sub-contractors, foremen, tradesmen and operatives than his colleagues who works for other architects.«

Speed in producing clear drawing, direct contacts without intermediaries, quick decisions and immediate instructions are essential. With this role, Segal believed himself much more integrated in the building team than any architect in a large practice, however »multi-professional«.

Relationship with clients

And how did he get work? »I am not aware of methods for acquiring clients and have never been officially recommended for a commission. People somehow have turned up and taken their chance with me.« (In fact for considerable periods Segal had very little work.)

But once he had a client, the first essential step was to set boundaries, to state the rules. And one of the clearest of these in this relationship was the demand that the client state his available resources honestly at the start. »*A client who does not commit himself to money from the start is dangerous*,« was a rule Segal

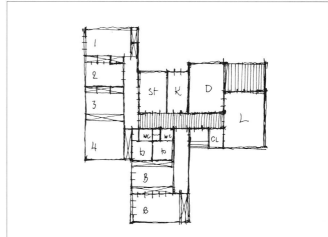

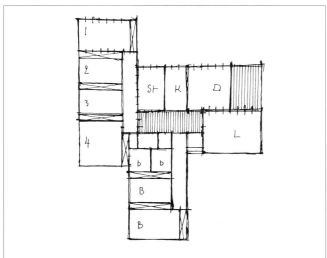

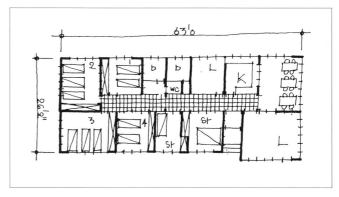

Various plans offered to the client for a children's home, Singleton, Sussex. They fell broadly into two groups, with staff accomodation either separate from or integrated with the children's rooms.
Mehrere Entwürfe für ein Kinderheim in Singleton, Sussex. Die Entwürfe fallen verschieden aus, je nachdem, ob die Personalräume in die Kinderzimmer integriert sind oder nicht.

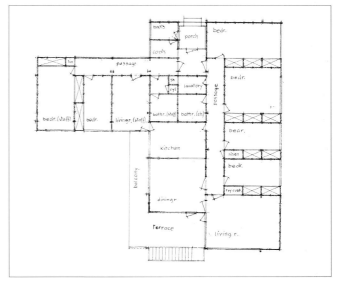

>
The chosen plan for Singleton.
Der schließlich ausgeführte Grundriß.

House M. R. Holland
School Lane, Bromeswell near Woodbridge, Suffolk
Calculation of Stresses in Timber Frame (based on CP 112)

Roofing Joist over widest span

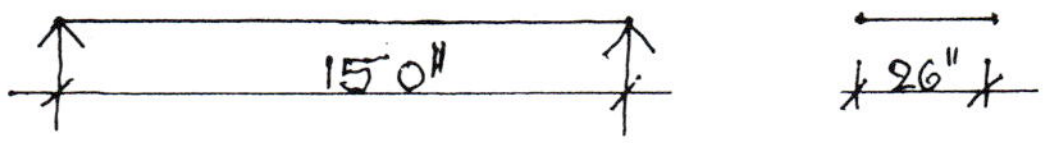

$$150''$$ $$26''$$

area to be supported
by joist : 26 sq ft

Loading

super load	26 sq. ft x 15 lb	=	390	lb
1" gravel	26 " x 9 "	=	234	"
2" wood wool slab	26 " x 4.5"	=	117	"
weight of joist	$\dfrac{2 \times 8'' \times 150'' \times 30\,lb}{1728}$	=	41.5	"

total   782.5 lb

permissible flexural stress (Group II timber) ≈ 800 lb sq in

$$M_B = \frac{782.5 \times 150''}{8} = 14700$$

$$M_R = \frac{14700}{800} = 18.3 \ \text{in}^3$$

assume $b = 2''$

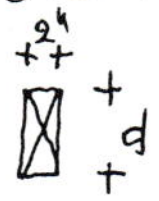

$$M_R = \frac{I}{n} = \frac{bd^3\,2}{12\,d} = \frac{bd^2}{6} = \frac{d^2}{3} = 18.3$$

$$d = \sqrt{3 \times 18.3} = \sqrt{54.9} = 7.35''$$

select 2" x 8" joist

A typical page of calculations.
Ein typisches Kalkulationsblatt.

hereinkam. Selbst heute noch fühle ich mich moralisch schuldig, wenn Moran (seine Frau) mich erwischt, wenn ich ein Buch lese, obwohl ich unter Druck bin ... Niemals hat mir jemand vorher gesagt, daß diese Routine vom Architekten verlangt würde.«

»Aber mit dem Selbstschutz, den ich mir aufgebaut habe, bleibe ich ungestört. Es gibt andere Dinge im Leben als nur ein Architekt zu sein. Die Ein-Mann-Firma braucht Zeit, sich zu entwickeln. Es ist auch nicht der Weg, der zum Reichtum führt oder ins Rampenlicht. Aber man gewinnt Freunde, die wiederkommen und – das ist das wichtigste – sie läßt Spielraum zur Muße und zum Leben.«

Er konnte diejenigen nicht verstehen, die sich ausbrannten in Arbeit, Hingabe und Ernsthaftigkeit. »Le Corbusier rannte zur Frühstückszeit, wenn ich im Bett bin, im Bois de Boulogne umher, dann malte er den ganzen Morgen hindurch, den ich mit Zeitunglesen verbringe, und nachmittags kehrte er ins Büro zur Architektur zurück.« Vom alten Aalto meinte er: »Wäre es für ihn nicht netter gewesen, wenn er weiterhin weniger und dafür bessere Bauten hätte produzieren können?«

»Das wichtigste Problem für einen Architekten ist, sowohl sein Ziel als auch seine Grenzen zu kennen, seine Ambitionen, das Ausmaß seines Talents, seine Gaben und seine Unzulänglichkeiten ins Gleichgewicht zu bringen. In meinem Fall wünschte ich über alles Freiheit in meiner Arbeit, und ich mußte lernen, nur soviel Arbeit anzunehmen, wie ich bewältigen konnte, ohne Verantwortung zu delegieren.«

»Wie funktioniert eine Ein-Mann-Firma? Die Antwort ist: Durch Vereinfachungen, Kürzungen, äußerste Konzentration und scharfe Beachtung der Wirtschaftlichkeit. Der alleinschaffende Architekt, der seine eigene Bauführung macht, arbeitet in viel engerem Kontakt mit dem ganzen Team: Das heißt mit den Baukostenspezialisten, Ingenieuren und Technikern, Unternehmern und Sub-Unternehmern, Vorarbeitern, Handwerkern und Arbeitern.« Klare Zeichnungen schnell herzustellen, direkte Kontakte ohne Vermittler, schnelle Entscheidungen und sofortige Anweisungen sind unerläßlich. In dieser Rolle fühlte sich Segal viel mehr dem Bau verbunden als ein Architekt in einem großen Büro, arbeite dieses auch noch so interdisziplinär.

Die Beziehung zum Bauherrn
Und wie kam er zu Aufträgen? »Ich weiß nichts von den Methoden, wie man Kunden aquiriert und bin niemals

durch jemanden direkt für einen Auftrag empfohlen worden. Die Leute sind einfach gekommen und haben ihr Glück mit mir versucht.«

Wenn ein Klient sich meldete, wurden zunächst die Grenzen abgesteckt, die Regeln bestimmt. Ganz wichtig war für Segal die Forderung, vom Bauherrn von Anfang an ehrlich über dessen finanzielle Möglichkeiten unterrichtet zu werden. *»Ein Klient, der sich nicht von Anfang an in der Geldfrage festlegt, ist gefährlich!«* Das war für Segal eine feste Regel. »Das Budget muß Teil des Auftrags an den Architekten sein. Verantwortungsbewußtes, auf Kenntnis der Verhältnisse beruhendes Geldausgeben ist untrennbar mit einem guten Entwurf verbunden.« Fehler im Programm führen zu Änderungen, und jede Änderung, sagt Segal, ist ein Zeichen professioneller Inkompetenz. In seinen Baukontrakten strich Segal die Klausel über Änderungen.

»Der Wohnbauarchitekt muß seinem Bauherrn mutig gegenübertreten, auf Entscheidungen bestehen und auf genügend Zeit, alle Zeichnungen vor Beginn des Baues fertigzustellen. Er muß seinem Kunden sagen, daß alle späteren Änderungen gefährlich und eine Quelle von Verzögerungen sind. Und er muß außerdem klarstellen, daß er mit der Schneiderin und Hutmacherin der Frau des Kunden nichts zu tun hat.« (Architectural Design, November 1953, S. 299, anonymer Leitartikel; Monica Pidgeon bestätigt, daß er von Walter Segal ist.)

Der Bauherr und sein Mitwirken am Entwurf
Der nächste Schritt ist »den Kunden die verschiedenen Möglichkeiten, wie sie sich aus dem Programm ergeben, nahe zu bringen, Prioritäten zu diskutieren und mit ihnen die optimale Lösung zu wählen.«

In der Regel werden dem Bauherrn zehn Pläne zu zwei oder drei prinzipiellen Lösungsvarianten gezeigt. Für jede Variante werden die Kosten der Konstruktion und Installation, Einsparungen bezüglich Lage und das Äußere untersucht. Mit Segals Holzbausystem können die Aufrisse dann schnell gezeichnet und der Baubehörde vorgelegt werden. Nach einer provisorischen Genehmigung kann dann sofort die vollständige Baueingabe mit Ansichten, Grundrissen und Lageplan, klar auf A4-Bögen gezeichnet, fertiggestellt und eingereicht werden.

Seit Mitte der Siebziger Jahre, seit Segal sich mehr und mehr in Selbstbauprojekten engagierte, entwickelte sich die Rolle des Bauherrn weiter; und zwar so, daß

underlined. »I cling firmly to the conviction that the budget is part of the architect's brief and that balanced, responsible and knowledgeable spending is inseparable from good design.«[2] Failure in the brief stage leads to variations; and every variation, said Segal, is a sign of professional incompetence. Indeed he was known to strike out the »variation order« clause on building contracts.

»To retain his small foothold, the housing architect must face the client courageously and insist upon decisions and upon time to complete *all* drawings *before* the work is started. He must also tell his client that variations are both dangerous and a source of delay. And he must further ensure that he should not be confused with Mrs Client's milliner or her dressmaker.«[3]

Client and design involvement

The next stage in the Segal practice is »to make my clients appreciate the variety of approaches to design solutions that arise from the brief, discuss priorities and select with them optimal solutions.« (See figs. p. 117.)

Typically, the client may be offered many plans, falling into two or three distinct patterns; each layout has explored economies offered by structural and service arrangements, by the location and aspect. Once a plan is chosen, with Segal's timber system, quick elevations can then be drawn and submitted immediately to the planning officer. With provisional planning approval, a full submission is made very quickly as elevations, plans and a site plan are submitted in clear outline on A4 sheets.

As Segal's work became more and more engaged with self-build from the mid 1970s, the role of client in designing also developed in parallel. So that by his last works Segal's clients were very much planning their own dwellings, using gridded paper and even models, within the precise rules of the architect's building system which we will look at shortly.

In earlier years, Segal's next step had been to produce a dossier of the complete building, with all the details. This is a file of freehand not-to-scale but fully dimensioned drawings. Then, with a quantity surveyor when necessary, this was made the first estimate from which grew the final bill of quantities and working drawings.

»The working drawings too are freehand, quite mechanical without any embellishments. All sizes are

given. Longhand lettering and dimensioning, large and legible. Figures are drawn with scale in left hand, drafting pen in the right, directly on tracing paper. There is no setting up, no rough, no tracing. *All this means speed.*« By the 1970s, Segal was doing everything on A4 paper, which became a file, a very simple book on the building site. The freehand and dimensioned drawings, calculations, quantities, specification and a copy of Segal's building manual, 20 pages of catalogue of parts and 9 pages on how to put them together, had become the total job documentation. The 9-page standard text, written in non-technical language, is a step-by-step manual for the constructor. Read in conjunction with the catalogue of elements, it sets out all the »how-to« principles of the Segal building method. The schedule replaces a traditional »bill of quantities« which aggreggates items with an accurate and detailed shopping list of materials required. Listed in order of erection, materials ordered and loaded from it are inevitably (and intentionally) sorted before delivery. Delivered to site and stacked, they are in the correct order for use. Such tiny points, packed with common-sense (particularly for small sites and inexperienced builders, but also avoiding double handling), abound.

This schedule, with each element detailed and located diagramatically, can then be used by the builder as a bunch of working details. When complete, the schedule is priced, and with each component itemised Segal can prepare an extremely accurate estimate for his client, which will not significantly vary.

The calculations exemplify Segal's method. Every structural member is calculated; a typical Segal frame has 11 or 12 A4 pages of calculations. Submitted alongside the drawings and specification for building consent, they delight authorities with their clarity and completeness.

To the end of his life, Segal calculated his own structural members and took his own quantities, which, as he admitted, »is so hellishly dull and repetitive.« In 1970 he commissioned a study[4] which showed that even then all the documentation for a typical project could be prepared by a correctly programmed computer within two days of receiving the layout line drawings.

At the time, before the proliferation of micro-computers, such a system ill-fitted normal working routine. Segal explained: »I didn't pursue it because of the cum-

page 3

LIST and QUANTITY of MATERIALS for ASSEMBLY KIT

PROJECT / ADDRESS / CLIENT : bungalow / School Lane, Bromeswell, Suffolk / Mr. & Mrs. M.R. Holland

material	description & location	grade	section or unit size	length & quantity or total area	finish	price £
framing timber	battens between coupled roof beams one face only to be planed	2	50 × 50	4/4000 1/4300	planed on one side only	
	facias to perimeter of floor, porch and terrace	Parana pine	1"×10" imp.	4/7'0" 2/9'0" 5/11'0" } no length 1/12'0" } shater 5/13'0"	prepared	
	facias to roof and terrace	do	1"×12" imp.	1/12'0" 1/13'0" 4/14'0" } no length 5/16'0" } shater 1/17'0" 1/20'0"	do	
	noggings to facias of roof	2	50×100 met. 25×25	1/3200 1/1800	do	
	fillets to roof	2	met	2/3100 1/3700 1/4000 2/4300 5/4900 1/5200	sawn do do do do do	
	cappings to roof	2	25×125 met	2/3100 1/3700 1/4000 2/4300 5/4900 1/5200	prepared	
	battens to secure cappings to facias	2	25×75 met.	2/3100 1/3700 1/4000 2/4300 5/4900 1/5200	do	
	blocks to battens as above	2	25×38	1/2400	do	
	wedges to hold down felt	2		1/2400	sawn	

A page of a list of quantities showing details.
Eine Seite aus einem Mengenauszug mit Detailzeichnungen.

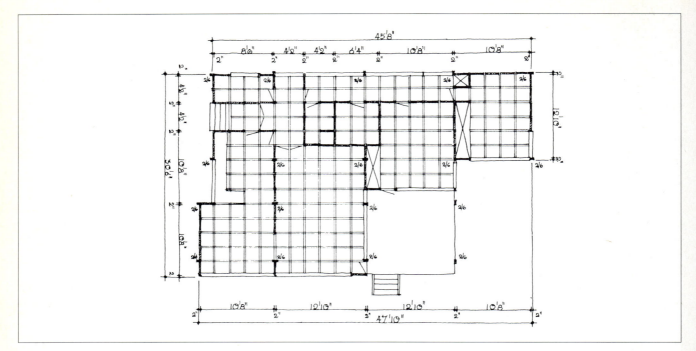

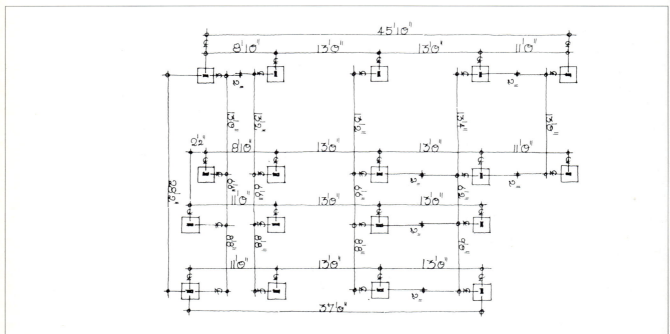

Typical set of Segal plans (Holland house, 1971): grid plan; setting out; flooring joist plan; roofing joist plan.

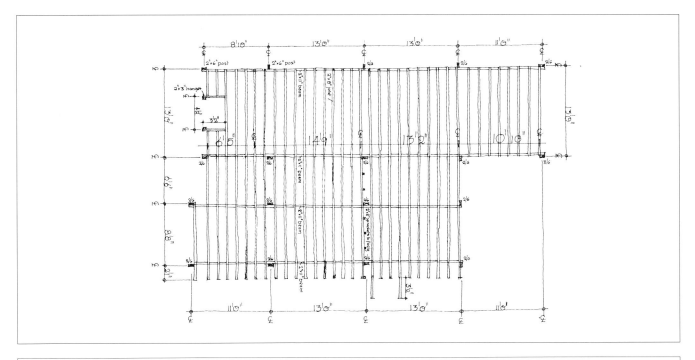

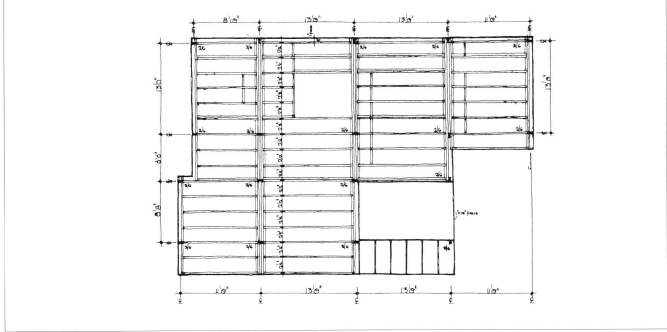

Typisches Set von Segal-Plänen, für das Haus Holland, 1971: Grundraster; Setzmuster der Stützen; Bodenbalken, Dachbalken.

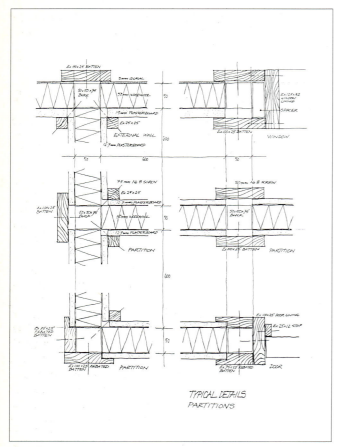

TYPICAL DETAILS
PARTITIONS

»Typical partition details«. A page from the Catalogue of Elements, drawn by Jon Broome.
»Typische Anschluß-Details«. Eine Seite aus dem Katalog der Bauelemente, gezeichnet von Jon Broome.

– which also allows me to intervene as I go, to vary the materials....« Of course today, with Fax and micros with ›modem‹, the aid of micro-electronics could be considerable. The point here is how Segal's way of looking at things always tried to consider the whole event, the humane processes in the production of homes, rather than being seduced by amazing efficiency in a little part of that process.

Later, structural engineers Ove Arup and Partners, also coded his »form of building«, allowing the varied documentation to be assembled very simply while leaving the choice of actual materials to the user to insert. With this system, Segal argued, no further »designer« was necessary. »New materials could easily be incorporated, and when the support structure, quantities, pricing, was all ready, you would be able to increase the range of buildings you can put up with such methods. In contrast to all the efforts to computerise drawings, this method, of course, completely dispenses with drawing.« Thus, he concluded, with the clients having inter-active contact with these tools and therefore (within the system) designing their own building, »the architect will no longer be necessary in that way, unless willing and able to take part in the further development of the structures.«

By 1970, Segal looked forward to »a time when the single highly-geared architect, powered by modern methods of documentation and administration provided by apparatus, will be in a convenient position to tackle the work load of the present day mammoth office using traditional methods. Then the re-individualised architect with power at his disposal and tools for his work (which consist of apparatus and not of management techniques) will be able to devote himself again fully to the job of designing in freedom: and in freedom it will be, as those that already have a taste of operating building methods of greater variability than traditional techniques know well.« His whole drive, as we have seen, had been to eliminate unproductive work and concentrate on designing.

But by the 1980s, however, his perspective had radically changed – or, put his way, taken the next logical step: if the system can so be set up to give architects space for designing, and designing within the system is made so comprehensible, then why cannot clients themselves take over this »re-individualised power«?

bersome problem of access. The computer was not here but at the City University. All the details and elements of my structure are coded. It would simply have meant, to get one of my typical bills of quantities, coding the card. This would take, shall we say, up to two hours. But then the process slowed down; I had to take it down by hand or send by post to the computer centre. There it would be printed out very quickly and would be returned to me by the same slow way, by post. Altogether, shall we say, taking four days. Now in four days I can produce, sitting here, a bill of quantities of my own

bei Segals letzten Arbeiten die Bauherren ihre Wohnungen weitgehend selber planten, mit Hilfe von Karopapier und sogar Modellen, im Rahmen der präzisen Regeln von Segals System, das wir später betrachten werden.

Nach der Plangenehmigung war Segals nächster Schritt, ein Dossier des ganzen Gebäudes mit allen Details anzulegen, einen Ordner mit freihändig gezeichneten, nicht maßstäblichen, jedoch voll vermaßten Zeichnungen. Dann wurden, wenn nötig mit einem Spezialisten, der Kostenvoranschlag, dann die endgültigen Mengenauszüge und Werkpläne gemacht.

»Die Werkpläne sind auch freihändig gezeichnet, ganz mechanisch, ohne Verschönerungen. Alle Maße sind angegeben; alles ohne Abkürzungen beschriftet und bemaßt, groß und leserlich. Alles gezeichnet mit dem Lineal in der linken, der Zeichenfeder in der rechten Hand, direkt auf Pauspapier. Es gibt kein Vorzeichnen, kein Nachziehen. All das bedeutet Schnelligkeit.«

Seit den Siebziger Jahren zeichnete Segal alle Pläne, von den ersten Skizzen bis zum Konstruktionsdetail, freihändig auf A4-Bögen. Berechnungen, Mengenauszüge und Ausschreibungen waren auf gleichen A4-Bögen. Dazu kamen eine Kopie von Segals Bau-Handbuch, 20 Seiten Katalog der Bauteile und 9 Seiten mit Angaben, wie sie zusammenzusetzen sind. Das war dann die gesamte Baudokumentation.

Segals Baubeschrieb ist ein neunseitiger Text in nicht-technischer Sprache, eine Schritt-für-Schritt Anweisung für den Erbauer. Wenn man sie zusammen mit dem Katalog der Bauelemente liest, werden alle Ausführungsdetails klar. Der Katalog der Bauteile ist gleichzeitig eine akkurate und detaillierte Einkaufsliste. Die Materialien sind in der Reihenfolge aufgeführt, in der sie gebraucht werden. Wenn sie entsprechend, nämlich in umgekehrter Reihenfolge geladen werden, sind sie schon vor der Lieferung sortiert. Wenn sie dann abgeladen und gestapelt sind, liegen sie in der richtigen Reihenfolge für den Bauprozeß bereit.

Die ganze Dokumentation, bei der jedes Element detailliert und im Plan diagrammatisch festgelegt ist, kann vom Handwerker wie ein Bündel von Detailzeichnungen benutzt werden. Wenn die Stückliste fertig, jede Komponente aufgeführt und das Ganze mit Preisen versehen ist, kann Segal dem Bauherrn einen außerordentlich genauen Kostenvoranschlag liefern, bei dem es keine wesentlichen Abweichungen mehr geben wird.

Auch die statischen Berechnungen sind beispielhaft für sein Vorgehen. Jedes tragende Bauteil ist berechnet. Zu einem typischen Holzständerbau von Segal gehören 11 oder 12 A4-Seiten Berechnungen. Wenn er diese zusammen mit Zeichnungen und Materialauszügen zur Genehmigung einreichte, waren die Beamten stets beglückt über die Klarheit und Vollständigkeit.

Bis zum Ende seines Lebens machte Segal seine statischen Berechnungen selber, wie auch die Materiallisten. Er gab zu: »Das ist höllisch langweilig und wiederholt sich ständig.« Schon 1970 gab er eine Studie in Auftrag (bei Nader Seil, City University, London), die belegte, daß schon damals die ganze Dokumentation für ein typisches Projekt von einem korrekt programmierten Computer in zwei Tagen, vom Eingang der Zeichnungen an gerechnet, hätte ausgeführt werden können.

Zu jener Zeit jedoch, vor der Verbreitung der Kleincomputer, war ein solches System schlecht in die normale Arbeitsroutine einzupassen. Segal erklärte: »Ich habe das System der Computerisierung nicht weiter verfolgt wegen des umständlichen Zugangs. Der Computer war nicht hier, sondern an der *City University*. Alle Details und Elemente meiner Konstruktion sind bereits codiert. Um also einen meiner typischen Mengenauszüge zu erhalten, hätte man bloß die Karte codieren müssen; das würde, sagen wir einmal, nur zwei Stunden dauern. Aber dann fing die Sache an, langsamer zu werden. Ich hätte die Karte entweder zum Computer-Zentrum bringen oder mit der Post schicken müssen. Das Ausdrucken wäre dann wieder sehr schnell gegangen, und dann hätte das Ganze auf demselben langsamen Weg, mit der Post, an mich zurückgehen müssen. Alles zusammen hätte das, sagen wir mal, vier Tage gebraucht. Und in vier Tagen kann ich, indem ich hier sitzen bleibe, selbst einen Mengenauszug herstellen, was noch den Vorteil hat, daß ich während der Arbeit eingreifen kann, zum Beispiel ein Material ändern ...« (Tonband/84)

Natürlich könnte heute, mit Telefax oder direkter Datenübermittlung, die Unterstützung durch die Elek-

Fig. pp. 126/7 / Abb. S. 126/7 >

Walter Segal and Jon Broome on site with a client/builder.
Walter Segal und Jon Broome mit einem Selbstbauer auf dem Bauplatz.

Administration

In the late 1960s Segal felt that »two of the most essential spheres in which the practicing architect invests a large amount of energy and time, and frequently most uneconomically so, are administration and documentation, and in both a veritable revolution seems almost at our doorstep.« His own adminstration was so minimised it virtually didn't exist at all. »It is unproductive and unavoidable. I do not glory in dictating letters and wasting time. *There is no secetary.* I receive letters in confirmation of instructions I give on the telephone during the pre-contract stage. When I write myself it is for legal purposes under the contract. I am not easily telephoned. I do not use an information library but ring the Building Centre, The Building Research Station or whoever else can inform me. Once a contract is signed, contact by letter or telephone ceases. Their place is taken by site meetings which I call and which must be attended by everybody concerned with the operations. I take the minutes which are typed and circulated by the general contractor. The site meeting decides all issues and there is *only one file for the job: the minutes' file.* Everybody is enjoined not to write or to telephone. Should that become necessary the blame lies with me.«

Where labour is used it is by sub-contract only. The client, *de jure*, became contractor, insuring his own site, letting the sub-contracts, and immediately saving himself an estimated 10% in preliminaries and 20% in profit, insurance and covers. *De facto*, of course, the architect was in control. The senior carpenter on site rang Segal every evening, and the site meeting to decide all points, as outlined above, was convened every ten days.

Professionalism

Segal had little patience with anyone who relied on given rules rather than using their own rational empiricism. »I have found one of the most difficult things in my life is to accept authority,« he said. Unfortunately he was never able to escape the tyrannies of building control regulation which he considered conservative and incompetent, and planning »aesthetic« controls which he considered an intolerable impertinence.

Throughout his career, Segal was exhorting architects to stand up and fight for their values in their own work. »We cultivate feelings of guilt. We have an inordinate social conscience: we think that in an imperfect world we have to strive for perfection. (Nobody else thinks that way.) And we take blame.« He recounted tales of architects attacked and rejected, right back to Phidias, left to die in a debtor's prison. Poelzig, his old teacher, set an example which Segal loved to quote: »With bravado and affrontery, at a meeting called to determine why a water tower he had erected, collapsed, instead of a defence he presented them with a fresh design for replacement.«

Segal, when first invited to give an RIBA lecture (in May 1977), addressed his audience: »We should take a different attitude to those things. I'll tell you a story. There was once a king who had a barber who, when he was shaving him one day, cut the king. Whereupon the king shouted ›That comes from drinking!‹. ›Yes your Majesty‹, said the barber, ›it makes the skin so rough‹. There are always two to play a game.«

Twenty five years earlier, he wrote: »Britain has so much talent in architecture. Let the architects approach their clients, not like Mark Twain's writer approaching his publisher on his knees and shaking his fist, but with courage, enthusiasm and the desire to lead.«[5]

Notes and Quotations

1. He was alone from when Alan Blanc left in 1952 until Jon Broome joined him in 1978 to work on Lewisham. Blanc tells, in conversation with the author in April 1988, how he split with Segal when the latter turned down a large commission partly because it contained an element of rehabilitation work which Segal was unwilling even to consider.
2. This comes from one of the many notes Walter Segal wrote anonymously for »Astragal«, *The Architects' Journal* gossip column; this from 31 October 1962.
3. Anonymous leader in *Architectural Design*, November 1953; Monica Pidgeon, letter to the author April 1988, confirmed that these leaders were written to commission, and that some, including this one, were by Segal.
4. By Nader Seil, as part of his MSc at City University, London, 1970.
5. as 3.

Quotations from Walter Segal in this section from: 1968(C); 1973(B)4; 1974(B)1; 1976(C)2; 1976(C)1; 1982(C); Wills & Yeo interview 1984; 1953(B); 1970(B)2; 1946(B)2; 1977(B)2.

tronik ganz erheblich besser sein. Doch geht es um die Art und Weise, wie Segal die Dinge betrachtete, immer versuchte, den ganzen Vorgang bei sich und unter Kontrolle zu halten und sich nicht blenden zu lassen von einer ungeheuren Effizienz in einem kleinen Teil des Prozesses.

Die Systemingenieure Ove Arup und Partner haben Segals Baumethode codiert, wobei das Einsetzen der zum Schluß gewählten Materialien dem Benutzer überlassen bleibt. »Neue Materialien könnten leicht eingeplant werden, und wenn die tragende Konstruktion, Mengen und Preise vorliegen, wäre es möglich, die Zahl von Gebäuden, die man mit solchen Methoden errichten könnte, zu erweitern. Im Gegensatz zu all den Bemühungen um Computer-Zeichnungen, käme man bei dieser Methode natürlich ganz ohne Zeichnungen aus.« Somit, folgerte Segal, wenn die Bauherrn dieses Instrumentarium beherrschen, könnten sie – innerhalb des Systems – ihre Gebäude selbst entwerfen. Dann wird der Architekt so wie bisher nicht länger gebraucht, es sei denn zur Weiterentwicklung der Konstruktionen und der Methode im Ganzen.

Um 1970 sah Segal die Entwicklung so: »Es wird die Zeit kommen, da ein einzelner, mit Apparaten gut ausgerüsteter Architekt in der angenehmen Position sein wird, die Arbeitslast eines heutigen mit traditionellen Methoden arbeitenden Mammut-Büros allein zu bewältigen. Dann wird sich der wieder zum Individuum gewordene Architekt ganz der Aufgabe, in Freiheit zu entwerfen, widmen können. Und eine Freiheit wird es wirklich sein, wie diejenigen schon wissen, die einen Versuch gemacht haben, mit Baumethoden von größerer Variabilität als den traditionellen Techniken zu arbeiten.«

Administration

»Zwei der wesentlichsten Bereiche, in die der Architekt eine große Menge von Zeit und Energie steckt – und in der Regel ganz unwirtschaftlich – sind Administration und die Dokumentation; und auf beiden Gebieten scheinen wir an der Schwelle einer wirklichen Revolution zu stehen«, sagte Segal in den späten Sechziger Jahren.

Walter Segals eigene Administration war minim, existierte fast nicht. »Es ist so unproduktiv und doch nicht vermeidbar. Ich genieße es nicht, Briefe zu diktieren und damit Zeit zu vergeuden. Ich habe keine Sekretärin. Ich bekomme Briefe als Bestätigung von Instruktionen, die ich am Telefon erteilt habe. Wenn ich selbst

schreibe, so aus rechtlichen Gründen im Rahmen des Kontraktes. Man kann nicht leicht mit mir telefonieren. Ich bediene mich für Informationen keiner Bibliothek, sondern rufe das Bauzentrum an oder die Bauforschungszentrale oder andere Informationsquellen. Nachdem die Kontrakte unterschrieben sind, gibt es nichts mehr zu schreiben oder telefonieren. Hingegen setze ich Treffen auf der Baustelle an, zu denen jeder erscheinen muß, der am Bauvorgang beteiligt ist. Ich führe Protokoll, das dann getippt und an alle verteilt wird. Bei diesen Treffen werden alle Punkte entschieden, und es gibt dafür nur eine Ablage, den Ordner mit den Protokollen. Jedem wird eingeprägt, weder zu schreiben noch zu telefonieren; sollte es trotzdem nötig werden, liegt es an ihm.«

Professionalismus

Durch seine ganze Laufbahn ermunterte Segal Architekten, für den Wert ihrer eigenen Arbeit einzustehen und zu kämpfen. »Wir kultivieren ein Schuldgefühl. Wir haben ein übermäßiges soziales Gewissen: Wir denken, daß wir in einer unperfekten Welt die Perfektion anstreben müssen. Niemand anders denkt so. Und wir nehmen die Schuld auf uns.« Er erzählte Geschichten von Architekten, die angegriffen und abgewiesen worden waren, bis zurück zu Phidias, den man in einem Gefängnis hatte sterben lassen. Poelzig, sein Lehrer, hatte ein Beispiel gesetzt: »Mit Bravado und Unverschämtheit trat er bei einem Treffen auf, das einberufen war, zu klären, warum ein Wasserturm, den er gebaut hatte, eingestürzt war. Statt sich zu verteidigen, präsentierte er einen frischen Entwurf für den Neubau.«

Bei einem Vortrag vor dem britischen Architektenverband (RIBA) im Mai 1977, wandte er sich an seine Zuhörer: »Wir sollten eine andere Einstellung zu den Dingen finden. Ich werde Ihnen eine Geschichte erzählen. Es gab einst einen König, der hatte einen Barbier, der den König eines Tages beim Rasieren schnitt. Darauf rief der König: ›Das kommt vom Trinken‹. ›Ja, Ihre Majestät‹, sagte der Barbier, ›das macht die Haut rauh‹. Es braucht immer zwei für ein Argument.«

Fünfundzwanzig Jahre früher schrieb Segal: »Es gibt in Großbritannien soviel Talente in der Architektur. Laßt den Architekten dem Klienten nicht gegenübertreten wie der Schriftsteller bei Mark Twain seinem Verleger: auf den Knien und gleichzeitig die Fäuste schüttelnd, sondern mit Mut, Begeisterung und dem Wunsch zu führen.«

»The Little House«

In his second and last lecture at the RIBA (in 1982) entitled »View from a Lifetime«, Segal talked about the 1920s, recounting his own development to »La Casa Piccola« of 1932. Without pausing for breath, he leaped thirty years, continuing: »About 18 years ago, I built in our garden a small structure...« and he opened into the housing forms of his later years. Somehow the 30-year career in between was too heavy to rehearse. (He often referred to it, unable to forget the central problems of traditional wet building contracting, as his »Thirty Years' War.«) Certainly his new »little house« both linked back to Casa Piccola and was a springboard for a newly directed practice.

The Segals, married in 1962 and with their large family, needed somewhere to live while the old house was demolished and replaced. It had to fit in the garden and be easily removed afterwards. »The money was to come out of the sum we had available for the

The Little House in summer, 25 years after construction (1963).
Das »kleine Haus« im Sommer, 25 Jahre nach Erstellung (1963).

Das »kleine Haus«

In einem Vortrag vor dem Royal Institute of British Architects unter dem Titel »Blick über eine Lebensspanne« sprach Segal 1982 über die Zwanziger Jahre und seine eigene Entwicklung bis zur *Casa Piccola* im Jahre 1932. Ohne eine Atempause übersprang er dann 30 Jahre und fuhr fort: »Vor etwa 18 Jahren baute ich in unserm Garten ein kleines Haus ...« Dieses neue *kleine Haus* schloß an die *Casa Piccola* an, bildete aber gleichzeitig den Anfang einer neuen Praxis.

Die Segals mußten 1962 mit der großen Familie irgendwo leben, während ihr altes Haus abgerissen und das neue aufgebaut wurde. Das Provisorium mußte in den Garten passen und später leicht wieder abzubrechen sein. »Das Geld dafür mußte von der Summe abgezweigt werden, die für das ganze Unternehmen zur Verfügung stand, und ich beschloß zu versuchen, was sich mit 700 Quadratfuß [ca. 65m²] zu einem englischen Pfund/Quadratfuß anfangen ließ.« Zum Schluß wurden

Segal watering the roof of his »Little House«.
Segal wässert das Dach des »kleinen Hauses«.

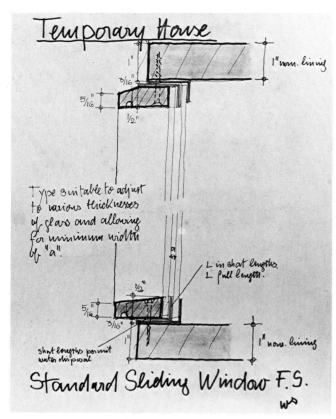

Sketch section of window.
Skizze eines Fensterdetails.

whole enterprise and I decided to see what about 700 sq.ft. and £1 per sq.ft. would give us.« In fact it cost £835 for a 715 sq.ft. (66.5m²) house (i.e. £12.55 per m²) which was built in ten weeks[1] and remains happily 25 years later.

Assembly not building

»The general concept,« Segal explained in 1971, »is that of Meccano, in which mass-produced materials are assembled in their market sizes. The framing which I used is a traditional one: a so-called ›balloon frame‹. In the whole context of this house the infil parts of the structure are in no way fixed together. When I started I decided to keep the resale value of the materials high in case the building was to be demounted. Thus none of the basic wall and partition materials have been touched by bolt, nail or screw: they are simply held in

position by friction. I slithered into the discovery, shamefully late, that a market of mass-produced material does exist, that, by and large, there are many materials that are dimensionally co-ordinated, and that my job was to think of *assembly* methods.«

The »Segal method« is not an invented »system«, but a design of assemblies. The little house was assembled from readily available materials in two shapes, basically boards and sticks. The sticks are planks of standard sawmill dimensions, 50 × 150mm for joists and 50 × 100mm for posts; the boards are 130 woodwool slabs used as roof, wall and partition, none of which is cut, bolted, screwed or nailed.

The plan is formed by a rectangular carpet of four rows of 13 slabs, which form a roof, sitting on joists and the loads carried to 20 posts at 1.85m centres. Each post stands on a flat 600 × 600 × 50mm paving slab on the ground (which slopes 1:7). There is no damp-proofing or foundation. No column takes more than 0.6ton load and none of the paving slabs has moved. The roof is flat, and simply covered with two layers bitumenous felt, lying on the substructure like a skin – but not attached to it. Scattered bricks and a 40mm sheet of water hold it down; the water level is maintained by hose, an overflow takes off particularly heavy rain.

The exterior wall surface is green mineral felt stuck to the slurried face of the slabs, and on the inside uncut sheets of 3mm hardboard (rough side out) is used, wedged under the window cills, hanging downwards. 50 × 50mm cover battens on both sides of the walls are bolted through, to hold the slabs in position, and an identical system is used on the partitions which are finished with waste chipboard paper, held behind the battens without adhesive. The floor is 25 × 150mm tongued-and-grooved boards with 25mm insulation underneath.

Window openings are as inventive as any other detail in this structure: they are simply lined in 25mm timber and then glass sheets slide in double aluminium angles. The outer angle, in short lengths with gaps, allows rainwater and condensation to enter the track and run to the outside, avoiding sealing problems. The window surround at sill, head and sides, typifying the Segal sense of »appropriate economy«, are of the best quality gurjun hardwood, ex 25 × 40mm and pared down to 8mm at their leading edge.

es dann 66,5m² für 835 £ (£12,55/m²). Das Haus wurde innerhalb von 10 Wochen gebaut und überstand 25 Jahre glücklich.

Zusammensetzen, nicht bauen

»Das generelle Konzept«, erklärte Segal 1971, »ist das des Meccano-Baukastens, der massenproduzierte Materialien in marktgängigen Abmessungen bereithält. Das tragende Gerüst war der traditionelle sogenannte *balloon frame*. Im Gesamtkonzept des Hauses sind die raumabschließenden Elemente nicht untereinander verbunden. Zu Beginn hatte ich beschlossen, daß der Wiederverkaufswert der Materialien so hoch wie möglich sein sollte, für den Fall, daß das Gebäude wieder abmontiert würde. Deshalb wurden Wand- und Trennwandelemente nicht durch Nägel oder Schrauben beschädigt, alles ist zusammengeklemmt. Ich machte spät, schändlich spät die Entdeckung, daß es einen Markt von massenproduzierten Materialien gibt, die in ihren Dimensionen aufeinander abgestimmt sind und daß es mein Job sei, an Verbindungsmethoden zu denken.«

Das *kleine Haus* wurde zusammengesetzt aus leicht erhältlichen Materialien, im Grunde in zwei Formen, Brettern und Stäben. Bei den Stäben handelt es sich um Planken, in den genormtem Abmessungen der Sägewerke, 50 × 150mm für Balken und 50 × 100mm für Pfosten. Die Holzwollplatten, die für Dach, Wände und Trennwände verwendet wurden, sind weder geschnitten, noch mit Bolzen, Schrauben oder Nägeln befestigt.

Der Grundriß beruht auf einem rechteckigen Raster von vier Reihen mit je 13 Platten, die das Dach bilden und auf Balken liegen, die ihre Last auf 20 Pfosten, Achsabstand ca. 2m, übertragen. Jeder Pfosten steht auf einer Bodenplatte, 600 × 600 × 50mm, direkt auf dem Boden (mit einem Gefälle von 1:7). Es gibt keine Fundamente und keine Feuchtigkeitssperren. Kein Pfosten hat mehr als 0,6 Tonnen Belastung aufzunehmen, und keine der Bodenplatten hat sich je bewegt. Das Dach ist flach und einfach mit zwei Lagen Dachpappe gedeckt, die wie eine Haut auf der Unterkonstruktion liegen, aber nicht auf ihr befestigt sind. Ziegelbrocken und 40mm Wasser halten sie unten. Der Wasserstand wird einfach mit einem Schlauch stabil gehalten; ein Überlauf läßt das Wasser bei starkem Regen abfließen.

Die Wände sind außen mit grüner Mineralpappe abgedeckt, die auf der rauhen Seite der Platten befestigt

The Little House: frame under construction in Segal's garden.
Das »kleine Haus«: Das Fachwerk des Hauses in Segals Garten.

ist, für die Innenseite wurden 3mm Hartholzplatten (rauhe Seite nach außen) verwendet, unter den Fensterbänken eingeklemmt, frei nach unten hängend. 50 × 50mm Deckleisten auf beiden Seiten halten, zusammengeschraubt, die Platten in ihrer Position. Die Trennwände sind genau so konstruiert, mit dem Abfallpapier der Spahnplatten tapeziert und durch die Decklatten, ohne Klebstoff gehalten. Der Fußboden besteht aus 25 × 150mm, mit Nut und Feder untereinander verbundenen Brettern, mit 25mm Isolierung darunter. Die Fensteröffnungen sind ebenso erfindungsreich wie jedes andere Detail an diesem Bau: Sie sind einfach mit 25mm Holzleisten eingefaßt, und die Glasscheiben gleiten in doppelten Aluminiumwinkeln. Die äußere Winkelschiene ist von Zeit zu Zeit unterbrochen, um Regen und Schwitzwasser nach außen abfließen zu lassen; damit entfallen alle Probleme einer Versiegelung. Diese Fenstereinfassungen, oben, unten und an den Seiten sind aus Gurjun-Hartholz allerbester Qualität, 25 × 40mm, zugeschnitten bis auf 8mm in den Führungsnuten.

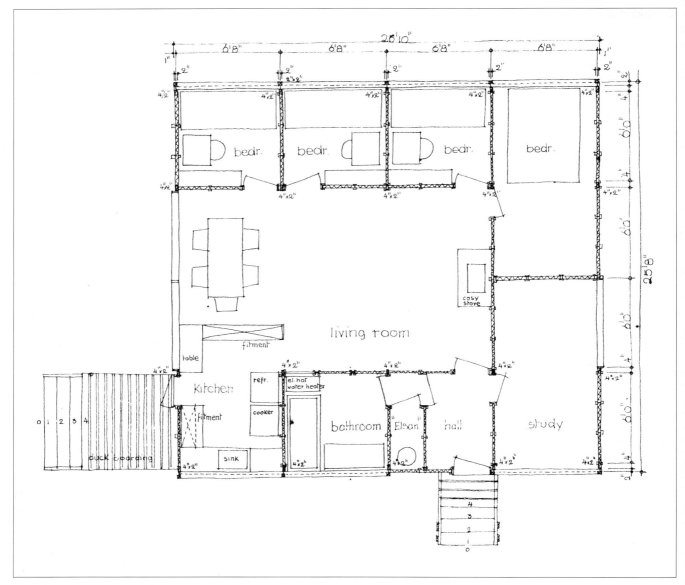

Plan of »The Little House«, dated 9/63.
Grundriß des »kleinen Hauses«, datiert 9/63.

The miniature »hall« house

»When the building was complete,« I quote again from Segal's lecture in 1971, »we started life in a house which was vastly below ›Parker Morris‹ standards[2]; but the living room was very much larger than anything in the standards, and this tilted the balance. For the children

small bedrooms each only 36 sq.ft. (3.77m²); our own bedroom, the size of a ›Parker Morris‹ single bedroom. My own office in the house was no larger. There was a kitchen, a bathroom and w.c. .«

Segal, who had spent so much time at the end of the Second World War working with tiny (often

Das Miniatur-Hallenhaus

»Nachdem das Haus fertig war«, ich zitiere hier wieder aus Segals Vortrag von 1971, »fingen wir an darin zu leben, in einem Haus, weit unter den Minimalstandards des sozialen Wohnungsbaus; aber der Wohnraum war viel größer als in den Standards, und das gab den Ausschlag. Die Kinderschlafzimmer waren sehr klein, je 3,77m²; unser eigenes Schlafzimmer entsprach etwa dem Standard für ein Einzelzimmer; mein Büro, auch im Haus, war nicht viel größer. Es gab eine Küche, Badezimmer und WC ...«

»So ein hierarchischer Grundriß mit Schlafzimmern und Nebenräumen rings um den Wohnraum ist praktisch, wenn man kleine Kinder hat, jedoch weniger gut, wenn die Kinder älter werden und anfangen, ihre eigene Persönlichkeit zu entwickeln. Dann braucht man etwas anderes; aber damals paßte er uns gut. Wenn meine Frau in der Küche zu tun hatte, waren die Zimmertüren der Kinder offen, sie konnten mit ihr sprechen und fühlten sich nicht allein. In so einem Haus reicht es, wenn man den zentralen Raum heizt und die Türen offen läßt; damit spart man viel. Alles war sehr primitiv, aber angenehm. Wir hatten Parties; es kamen Freunde; es war ein billiges Haus und billig zu bewirtschaften. Die Erfahrung mit bezahlbaren und durchaus nicht bedrückenden Lebensbedingungen läßt Einmischung besonders verdrießlich erscheinen: Man erzählte uns, wie wir unser Geld ausgeben sollen, weil man zu wissen meinte, was gut für uns sei.«

Der Grundriß zeigt die Gliederung in »Wirtschaftshaus«, »Kinderhaus« und »Elternhaus«, alles um die zentrale Halle. Dieses war der Haustyp, den Segal 1934 auf den Balearen analysiert hatte: Wirtschafts- und Schlafräume rings um die Sala, den Familienraum.

Es ist alles so winzig, daß man es als Modell oder Prototyp betrachten muß. Der Plan steht einerseits in Beziehung zu Segals Essay über Ibiza, *The Traditional Mediterranean House*, hat andererseits ein Vorbild: Hinten in Konrad Wachmanns *Holzhausbau*, das den Studenten Segal so sehr beeindruckte, ist ein hübsches kleines Gebäude, ein bescheidenes Ferienhaus in Holzbauweise von Hans Poelzig abgebildet. Es ist fast gleich groß, ebenso direkt und einfach im Material und der räumlichen Ordnung.

Auf Segals Zeichnung steht: »Wieder aufzubauen als Ferienhaus.« Vielleicht war das der Grund für die Baubehörde, daß sie nur eine zeitlich befristete Baugenehmigung erteilte. Aber als die Zeit des Abbruchs kam, hatte seine Frau so viele Einwände, daß Segal einen langen Kampf mit den Behörden aufnahm, damit es stehen bleiben konnte.

Viele Jahre später erinnerte sich Moran Segal der Schwierigkeiten, auf so engem Raum mit heranwachsenden Kindern fertig zu werden, besonders wegen der unumgänglichen strengen Regeln über den Lärm. Damals jedoch, im Juli 1966, sagte Walter Segal: »Es wird schmerzlich sein, das kleine Haus zu verlassen, in dem wir so lang glücklich waren, und man fragt sich, warum wir überhaupt ein traditionelles Haus bauen müssen; das kann nur beantwortet werden von jenen, die weiser und besser sind als wir.«

Er war es müde, Gebäude in traditioneller Bauweise aufzustellen, und mehr noch hatte er genug von den Kämpfen, die es brauchte, sie nach den üblichen Methoden zu bauen und dann noch Jahre mit der Abrechnung zu verbringen. Anders als ihr festes Haus, erregte dieses provisorische Haus sofort Interesse in Fachkreisen (durch Colin Boyne in *The Architects' Journal*), aber auch in der breiten Öffentlichkeit. *Financial Times*, *Daily Mail* und *Ideal Home* brachten 1966 Artikel darüber. Kunden wünschten sich Häuser wie dieses, und Segal hatte von Neuem seinen Dreißigjährigen Krieg zu beginnen.

Corner of »The Little House«.
Ecke des »kleinen Hauses«.

»prefab«) house plans in efforts to make them as humane as possible, built confidently on that background.

»Such a hierarchic plan of house with bedrooms etc. arranged around the living space is practical when you have young children; it is not so good when your children get older and individualise themselves. Then, of course, you need a different plan; but at the time it suited us very well. It meant that while my wife was working in the kitchen, the children had the doors to their own rooms open and could talk to her and thus did not feel isolated. We spent a very small amount on heating; in such a house plan you can do this by heating the central room, and then leave the doors to the bedrooms open. All very primitive, but it was very pleasant. We had parties in the house, we had friends there; it was a cheap house and cheap to run. The experience of achievable and less onerous living conditions makes one resentful towards controls and planning interference with your life: they tell you how you should spend your money, because they think they ›know‹ what is good for you.«

The layout showed Segal's division of the house into a number of component little »houses«: here the service house opposite the children's, and the parents' house at the back, all round the central »hall« space. This was the house type Segal had analysed in the Baleares in 1934: the service and sleeping areas grouped round the »sala«, the family room. And over the next decades he made sketch designs for such houses.

But here it is so tiny that this can really only be taken metaphorically; for at this scale and with such lightweight construction, there are many problems centred on lack of space for individual privacies. It is a model rather than a prototype for relaxed family life.

While the plan links to what Walter Segal (in his essay on Ibiza) called »the traditional Mediterranean house«, it also displays another much quieter but appropriate precedent: that modest timber-frame holiday house by Hans Poelzig, hidden at the back of Konrad Wachsmann's *Holzhausbau* which so impressed the student Segal. This lovely little building is almost exactly the same area (a square of side 7.8m and Segal's 8.1m), it shares a direct and simple material and spacial ordering, and its »little houses« in a little courtyard house plan.

Segal's drawing states »to be re-erected as a holiday house«. Perhaps this was added for the authorities, who only granted it a temporary building licence. But when the time came to demolish, Segal claimed that Moran so objected, that he found it easier to take on the planning authorities in a long battle to let it remain.

Moran Segal, years later, recalled the difficulty of dealing with growing children in such a tight space, particularly the need for strict rules about noise. »Needless to say the sound problems were horrific – almost like living in a tent.«[3] During this period, John Segal was working for ›A. levels‹ with so little privacy that, many years later, he would remark, »...for the parents this was probably easier than for the children as the rigorously enforced regime was their own!«[4]

At the time, however, in July 1966, Walter Segal said, »It will be a wrench to leave the little house where we have been so happy for now nearly one year, and the question why we have to build a traditional house at all arises, and can only be answered by those who are wiser and better than we.«

He was »sick and tired of putting up buildings in traditional construction: more than sick and tired of fighting to get them built, using the standard method of procedure, and spending years thereafter to finalise accounts.« Unlike their permanent house, this little one generated immediate interest[5]; clients wanted houses like it, and Segal never had to reopen his »Thirty Years War«.

Notes and Quotations

1. It could, he said, have easily been built in five, but for insufficient power tools, awkward access, and double handling of all goods which had to come through the basement of the old house.
2. Minimum space standards for dwellings accepted by the UK Government at the time as mandatory in public housing; from a report chaired by Parker Morris, *Homes for Today and Tomorrow*, HMSO, 1960.
3. Moran Segal in conversation with the author, March 1988 and letter to the author, 9 June 1988.
4. Letter to the author, 27 August 1988.
5. Professionally through the encouragement of Colin Boyne at *The Architects' Journal* but also by articles in the *Financial Times, Daily Mail* and *Ideal Home* during 1966.

Quotations from Walter Segal in this section from: 1982(B); 1971(B)2; Wills & Yeo interview 1984; letter in *AJ* 13 July 1966.

Internal view of »The Little House« in the 1980s, with ceiling added and internal partitions removed.
Das Innere des »kleinen Hauses« in den 80er Jahren, mit neu eingezogener Decke und ohne Zwischenwände.

»The Little House«. First winter 1963/64.
Das »kleine Haus« in seinem ersten Winter, 1963/64.

Construction

Segal's constructional understanding grew from a remarkable sense, which colleagues noted thoughout his life, which was his innate feel for structure and instinct for building. As Florian Beigel said, »He related building to his bodily experiences. He understood the basic structural mechanisms which members pull, bend and push in an almost physical way.«[1]

This common-sense based on bodily experience is exemplified by his understanding of insulation and condensation. »It is perfectly obvious,« Segal would say, »that if you join materials of different densities together where they cannot ›breathe‹, you get condensation. Regulations today demand that modern buildings are properly sealed and insulated and so you get the new English disease: condensation!«

He neatly used the analogy of clothing: »If you go outside, you will be wearing a raincoat or something waterproof to keep the rain off; you will be having something to keep you warm; and then something to stop them scratching you, like a shirt. If you stick all these parts together in one single garment, you will be sweating immensely. There is going to be on the inside of the raincoat a lot of condensation; if it cannot get out it will move inwards, make your pullover damp and from there travel further. Likewise, you are making steam – a pint of moisture a day – and this too has to get out....

»In houses in the past there were open grates and high levels of ventilation, there were absorbing materials with a loose molecular structure like bricks. Today with walls of different materials, you must conduct the moisture down, let it out somehow.«

This understanding of moisture and movement is evident throughout Segal's career, and well seen in the prototype aluminium frame house he designed in 1949 (ABSE house). With his later timber frames, the layers of Segal's composite walls, only held together by friction, allow moisture to find its way out. The system automatically provides weep holes every 600mm. But, as Segal said:

»This was not designed! I didn't understand this in the beginning. But my houses were very dry; even in a very damp period in Ireland, my client was astonished to come home after two weeks away and find it so dry. I couldn't explain this; I rang the Building Research Station and they couldn't either. Then one day, standing with my back against the wall, I put my hand underneath the ›Glasal‹ sheeting and felt these weep holes – they were just there as an inevitable part of the construction.«

This story is recounted to show how, with such a coherent attitude to making building, in which each step has a transparent logic, the product ›automatically‹ works in ways other than those expected. It stands as a converse to »iatrogenic« architecture[2], that widespread phenomenon today which can be defined as building problems which, but for the intervention of the architect, would not have occurred and needed solving.

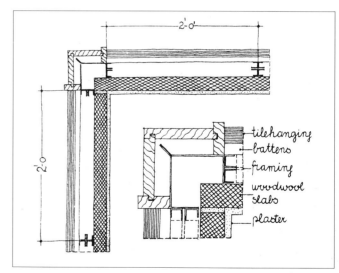

tilehanging
battens
framing
woodwool slabs
plaster

Plan detail of the »ABSE« aluminium frame house which Segal designed in 1949, showing composite wall.
Detail des »ABSE« Aluminium-Fachwerkhaus, das Segal 1949 entwarf: Wandaufbau.

Konstruktion

Segals Sinn für Konstruktionen haben seine Kollegen stets bewundert. Florian Beigel sagte: »Er brachte das Bauen in Beziehung zu seinen körperlichen Erfahrungen; er verstand die grundlegenden strukturellen Mechanismen, welche Bestandteile ziehen, sich biegen oder schieben, in einer fast physischen Weise.« (The Architects' Journal 4. 5. 1988)

Dieser auf körperlichen Erfahrungen beruhende Verstand kommt zum Beispiel in Segals Verständnis für Isolierung und Kondensierung zum Ausdruck. »Es ist doch offensichtlich: Sobald Ihr Materialien von unterschiedlicher Dichte zusammenfügt, entsteht Kondenswasser. Die heutigen Bauvorschriften verlangen, daß moderne Bauten ordentlich versiegelt und isoliert sind, und damit bekommt Ihr die neue Englische Krankheit: Kondenswasser.« Er erklärte es anhand der Kleidung:

»Wenn Ihr nach draußen geht, werdet Ihr einen Regenmantel oder etwas Wasserdichtes tragen, um den Regen abzuhalten; Ihr werdet auch etwas anziehen, das Euch warm hält, und dann etwas, damit es nicht kratzt, ein Hemd zum Beispiel. Wenn man jedoch alle diese Teile zu einem einzigen Kleidungsstück zusammenfaßte, würdet Ihr gewaltig schwitzen. Auf der Innenseite des Regenmantels gibt es eine Menge Schwitzwasser; wenn es nicht hinaus kann, wird es nach innen dringen, den Pullover durchfeuchten, dann das Hemd und so weiter. Außerdem gebt Ihr selbst Dampf ab, etwa einen halben Liter Flüssigkeit am Tag, und das muß auch nach außen gelangen ...«

»Früher gab es in den Häusern offene Kamine und einen hohen Grad an Ventilation; es gab absorbierende Materialien wie Ziegel. Heute, mit den

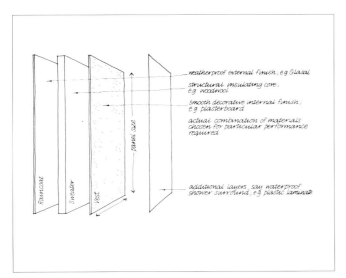

»Segal method«: Composite wall of timber Buildings. Nach Segal-Methode aufgebaute Wand für Holzbauten.

Prefabrication was first based on architect-designed components. This is different from Segal's later thinking (drawing from »Homes for the People« 1945, co-authored by Segal.)
Vorfabrikation beruhte zuerst auf Elementen, die Architekten entworfen hatten. Segal sah dies später anders. (Zeichnung aus »Homes for the People« 1945, dessen Mitautor Segal war.)

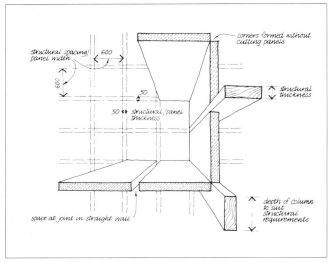

The modular system. Explanatory drawings by Jon Broome   (p. 140-145).
Das Modelsystem. Erläuternde Zeichnungen von Jon Broome (S. 140-145)

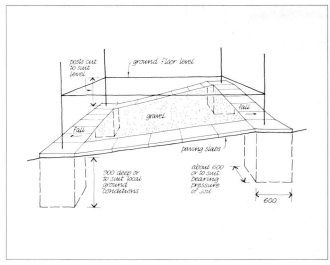

Foundations.
Fundamente

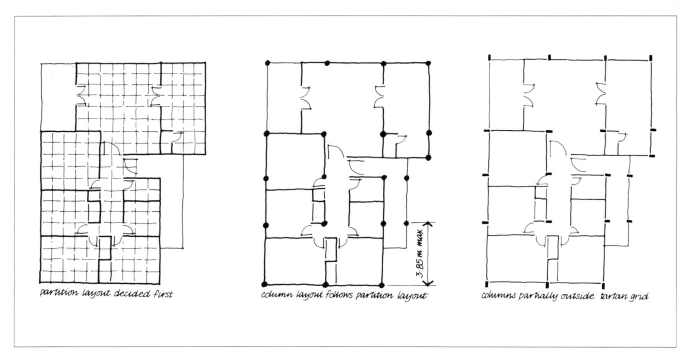

The partition layout is decided first. Next, the columns are laid out. Then the columns are placed, where possible, outside rooms.
Zuerst wird die Wandeinteilung bestimmt. Dann wird die Position der Stützen festgesetzt. Anschließend werden die Stützen plaziert, wenn möglich außerhalb der Räume.

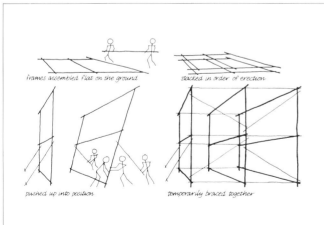

frames assembled flat on the ground

stacked in order of erection

pushed up into position

temporarily braced together

**Drawings to explain erecting of frames, 1966/67.
Schemazeichnungen, die das Verfahren erklären, wie
die Stützen aufzustellen sind, 1966/67.**

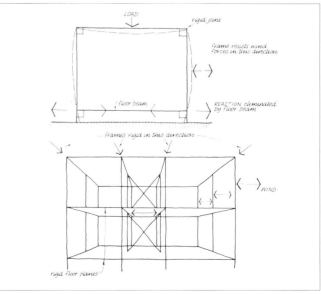

LOAD

rigid joint

frame resists wind
forces in this direction

floor beam

REACTION eliminated
by floor beam

frames rigid in this direction

WIND

rigid floor planes

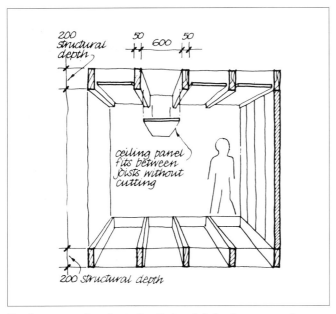

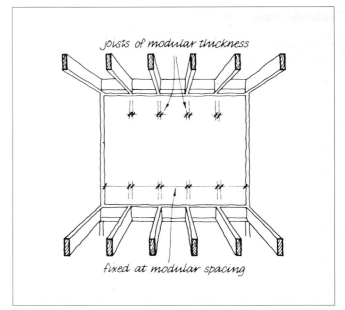

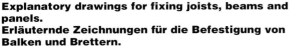

Explanatory drawings for fixing joists, beams and panels.
Erläuternde Zeichnungen für die Befestigung von Balken und Brettern.

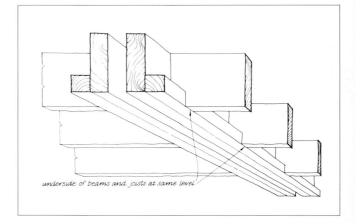

Segal was distressed when asked if he arrived at his method by trial and error. »Absolutely not,« he replied. »I have never had time for old ideas; I have always been *hyper*critical of traditional construction. But I have never designed by trial and error. I felt I had enough building experience – after all 28 years building things – to make it succeed. I have learned things that are not at all experimental; it was just a matter of common-sense.« The Segal method, then, is an attitude of mind; but one which quickly moved from common-sense construction to the possibility of a uniquely client-responsive housing. The developed method, beginning at »the little house« and refined over the decades, shows this sophisticated intermeshing of sensible ideas.

First, the house is planned. With building blocks of 600mm panels and 50mm wide posts, a »tartan« grid is set up which allows clients with little practice and assistance to develop layouts quite simply. As no walls bear loads, doors and windows can be located anywhere, indeed they can be moved during construction to a better position. Then, on this basic layout, the structure is worked out.

The roof is now on and the flooring being completed.
Das Dach ist gedeckt. Nun werden die Böden gelegt.

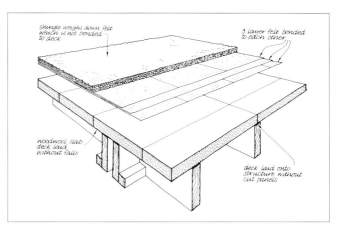

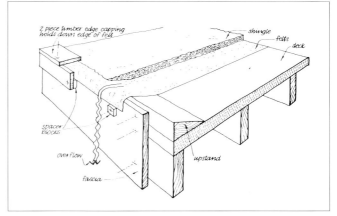

Each roofing layer is laid loose.
Jede Schicht des Dachaufbaus wird lose aufgelegt.

Wänden aus verschiedenen Baustoffen, muß man die Feuchtigkeit nach unten führen, sie irgendwie herauslassen.«

Segal war immer bekümmert, wenn man ihn fragte, ob er zu seinen Lösungen über Versuche käme. »Absolut nicht«, antwortete er, »ich habe niemals Zeit verbraucht mit alten Ideen; ich bin immer hyperkritisch gegenüber traditionellen Konstruktionen gewesen; aber ich habe niemals aufgrund von Versuchen konstruiert. Ich glaube, ich hatte genug Bauerfahrung – nach 28 Jahren –, um Erfolg mit meinen Konstruktionen zu haben. Ich habe Dinge gelernt, die ganz und gar nicht experimentell sind; es war alles eine Frage des gesunden Menschenverstandes.«

Zuerst wird der Grundriß geplant, auf der Basis eines Rasters, gegeben durch 50mm breite Pfosten in 600mm Abstand. Auch Bauherren mit wenig Praxis können darauf mit ein bißchen Hilfe einen Grundriß entwickeln. Da die Wände nicht tragend sind, können Türen und Fenster beliebig angeordnet, sie können sogar noch während des Bauens in eine günstigere Position verlegt werden. Dann, aufgrund dieses ersten Plans, wird die Konstruktion ausgearbeitet.

»Man macht also als erstes den Grundriß«, erklärte Segal. »Das ist ein großer Unterschied zu früheren Generationen des Skelettbaus, als man zuerst das Skelett plante, und dann den Grundriß hineinbrachte.« Im Gegensatz zu Corbusiers *plan libre* ist es ein Grundriß, der wirklich frei ist. »Ihr entwerft einen Grundriß aufgrund des Rasters, ich entwerfe eine tragende Kon-

»So what you do first is *plan*,« Segal explained. »Which is quite different from other earlier generations of framing where you design the frame first and then put a plan into it. *You* design a plan on a modular basis; and then I design a frame support structure. I look for beam lines, append supports to the beam lines, and then begin to calculate it and it sorts itself out.« (See figs. pp. 122/23, and 140 *bottom*.)

It neatly reverses the *plan libre* of Le Corbusier with a plan which really is free. In contrast to most timber building in Britain of stud and panel, the Segal frame is an expensive structure in material; usually stress-graded, knot-free Siberian or Canadian Douglas Fir, calculated to use minimum section. But it is a flexible tool, easily adaptable to the chosen plan, with very low foundation costs, and a clear span of six modules (3.85m) without deflection or, used most economically with an additional cantilevered module at each end, a possible clear space of over 5m.

Once the structure is calculated for the loads, deflection, bending and so on the frame layout may need amendment. Segal used cantilevers where possible, which could imply freestanding posts; otherwise columns are planned to project beyond the grid outside the building or into corridors (see fig. p. 140 *bottom*).

»Basically the concept is very simple portal frames, bolted together and joined (when two-storey) by St Andrews' Crosses developed – as in the Norwegian tradition – as compression members.« One subtle difference from a portal is the floor beam which, tying it together, eliminates horizontal reactions at the base (see fig. p. 141 *bottom right*). This is essential as the frames just stand, on a lead pad which closes the end-grain, unfixed on paving slabs. Regulations normally require a foundation pier about 900mm deep under each 600mm x 600mm slab. With such limited excavation, this can be simply hand-dug and filled with concrete from a wheelbarrow (see fig. p. 204 *bottom*).

Foundations are simply at whatever ground level exists, and posts are cut to suit. Thus Segal buildings can respond easily to »awkward« sites and can safely be near trees. Also, with frames not attached to foundations, normal fine tolerances for positioning and levelling foundations are unnecessary. Frames, assembled rigidly on the ground, are stacked in order of erection and then raised and braced, one by one (see fig. p. 141 *bottom left*).

Segal often compared his building with medieval practice, from the frame and construction techniques to the wall panels being infil between the posts. Galvanised steel bolts forming the rigid joints rather than timber pegs holding a morticed and tenoned connection form the only difference. Segal's bolt sizes are determined by the load on that joint and the timber used, which in turn determines the spacing and precise location of the bolts.

Floor and roof joists, of same section but much less strong timber, are then fixed on the module between the frames, and thus very quickly the roof can then be assembled (see figs. p. 142 *right*).

The deck of woodwool slabs is laid loose across the joists, so dimensioned to avoid any cutting. Then laid level over them, without falls, and loose without bonding, is the waterproof »tablecloth«. This (usually just three-layer felt) is folded over the edge and wedged down by a capping piece which simply presses on the felt. The roof has a generous overhang, which gives protection; for example it allows bedroom windows to be kept open at night. As Jon Broome says, »One effect is to create almost a microclimate of shelter immediately around the building.«[3] (See figs. p. 143 *right*.)

Because of evaporation and the awkwardness of topping up a second-storey roof, the water layer at Highgate is replaced with 40mm shingle. This holds down the tablecloth (and the building below) while also keeping it cool and damp. There is one rainwater outlet with a slight upstand, which comes through the overhang and is the only penetration in the covering; if it blocks, water can overflow the roof edge without any problem.

The floors of simple tongued-and-grooved boards are laid next, with insulation on panels under the lowest. There is then a comfortable working platform for assembling the wall.

The material of the wall sandwich can vary, but typically the matrix which gives substance and some insulation, Segal's »pullover«, is uncut 50mm woodwool slab while the »raincoat« is »Glasal«, a compressed fibre-cement panel, with a coloured enamel finish. »Quality products of this kind I never could use in previous days,« Segal notes, »when I had to follow the practice of building expensive carcasses and use cheap and short-life finishes.« Typically the inner surface, the »vest«, is uncut sheets of plasterboard, to which can be layered pinboard or waterproof laminate, for example, for study or bathroom. Extra layers, a

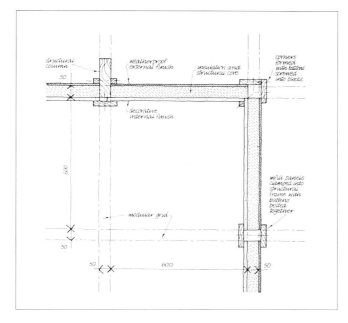

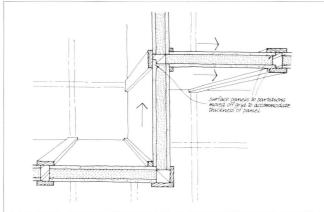

struktion. Ich finde die günstige Lage für die Balken, dann die Stützen, und dann fange ich an zu rechnen, und der Rest ergibt sich daraus.«

Im Gegensatz zu den meisten Holzbauten in Großbritannien, ist Segals Rahmen eine kostspielige Konstruktion. Er besteht aus astfreiem, in der Regel auf Zug beanspruchbarem, sibirischem oder kanadischem Douglas-Fichtenholz, das auf minimale Querschnitte berechnet wird. Aber es ist ein flexibles Werkzeug mit sehr geringen Fundamentierungskosten, einer freien Spannweite bis zu sechs Modulen (3,85m) ohne Durchbiegung.

»Im Grunde beruht das Konzept auf sehr einfachen Portalrahmen, in den Gelenkpunkten durch Bolzen verbunden, und, falls zweistöckig, durch St. Andreas-Kreuze ausgesteift, die, der norwegischen Tradition entsprechend, auf Druck gerechnet sind.« Im Unterschied zum Portalrahmen zieht der Fußbodenbalken die Konstruktion zusammen und schaltet horizontale Bewegungen am Boden aus. Das ist wesentlich; denn der Rahmen steht unbefestigt auf 600 x 600 mm Bodenplatten. Die Baunormen verlangen in der Regel ein 900mm tiefes Punktfundament unter jeder Bodenplatte. So etwas kann man von Hand ausheben und einfach mit Beton aus einem Schubkarren auffüllen.

Die Fundamentierung ist also einfach, egal wie das Geländeprofil aussieht. Die Pfosten werden auf entsprechende Längen geschnitten. Segal-Bauten können sich problemlos ungünstigen Grundstücken anpassen, auch bedenkenlos dicht neben Bäume gestellt werden. Da der Rahmen auf den Fundamenten nicht befestigt ist, entfällt auch die Feinnivellierung der Fundamente. Die Rahmen werden auf dem Boden fertiggestellt, in der Reihenfolge des Bauvorganges gestapelt und dann, einer nach dem andern, aufgerichtet und ausgesteift.

Segal verglich sein System oft mit mittelalterlichen Praktiken. Der einzige Unterschied besteht in den verzinkten Bolzen statt Holzpflöcken als Holzverbindung.

Details: External corner; internal partitioning; steel bolt clamps holding between them the composite outer wall (looking upwards).
Details: Aeußere Ecke; Innenteilung; Stahlbolzen, welche die Komponenten der Außenwand zusammenhalten (von unten gesehen).

The neat, compact suspended stair in Segal's own house.
Die zierliche, kompakte Treppe in Segals eigenem Haus.

Suspended, self-built stair in Lewisham.
Selbstgebaute, aufgehängte Treppe in Lewisham.

»shirt« of urethane foam for example between vest and pullover, will improve the thermal performance. (It is a construction which intrinsically leads to ventilation heat losses higher than those in conventionally sealed buildings.) Though the thermal performance is improved from Highgate, Jon Broome makes the important point of appropriateness: »Segal understood very well that there comes a point beyond which further elaboration negates the real advantages of simplicity.«[4] (See fig. p. 145 *top*.)

The window, simply refined from Highgate, is also not airtight, but it can be built as double windows or, in fact, various conventional windows can easily be fitted. The Segal window, very adaptable to different sizes of opening and fixed panes, remains an exemplary essay

in doing more with less: the two aluminium angles, spaced with a washer to let moisture out, hold sheets of glass with polished edges, which simply need a rub with a wax candle on the running edge before the pane is fixed in place. Simple charm – and a fine feeling of spaciousness from inside (see figs. pp. 159, 161).

Partitions are formed exactly as the external walls, and uncut plasterboard sheets laid loose on battens between the joists for the usual ceiling. All services, of course, can simply be accomodated in the easily accessible floor, wall and ceiling voids (see fig. p. 145 *middle*).

Segal always loved staircases, which througout his career illustrate his lively ingenuity; here is a chance to display little virtuosic pirouettes of joinery in minimal spaces. Within this building form, the stairs as separate

Other Lewisham self-built stair.
Eine weitere selbstgebaute Treppe in Lewisham.

Client-built shelf unit dividing kitchen from dining room.
Vom Klienten selbstgebauter Schaft als Trennung zwischen Küche und Eßzimmer.

Die Größe von Segals Bolzen wird von der Last bestimmt, die die Verbindung aufzunehmen hat und der Holzqualität; diese andrerseits bestimmt die Abstände und präzise Position der Bolzen.

Boden- und Dachbalken, im selben Querschnitt, aber aus wesentlich schwächerem Holz, werden dann dem Modul entsprechend zwischen den Portalrahmen befestigt. Damit kann das Dach schnell geschlossen werden. Eine Lage Holzwollplatten wird lose auf die Dachbalken gelegt, und zwar so dimensioniert, daß sie nicht beschnitten werden müssen. Darauf kommt dann, eben, ohne Gefälle, lose, unbefestigt das wasserdichte »Tischtuch«, in der Regel drei Lagen Dachpappe. Dieses wird über die Kanten gezogen, mit einer Abschlußleiste verkeilt, die einfach auf die Dachpappe

gepreßt wird. Das Dach hat einen großzügigen Vorsprung, der genügend Schutz gibt, um die Schlafzimmerfenster nachts offen zu lassen. Wie Jon Broome sagte, »eine Maßnahme, die fast ein geschütztes Mikroklima rings um den Bau erzeugt.« (Jon Broome: The Segal Method, Sonderdruck The Architects' Journal 5. 11. 1986)

Als nächstes wird dann der Fußboden verlegt, einfache Bretter mit einer Nut- und Feder-Verbindung und einer Lage Isolierplatten darunter. Damit hat man eine bequeme Arbeitsplattform für die Montage der Wände.

Die Materialien der Verbundelemente für die Wand können verschieden sein. Typisch ist ein Kern, der Substanz und eine gewisse Isolierung gibt,

treads can be supported on hangers, from posts, or hung from cantilevers. And the basic structure can easily be adapted to form pergolas, porches, verandahs or seats (see fig. p. 151). As Segal said, »From the roof joists you can hang anything – bay windows, for example, 600mm (one module) wide, increase the width of the house wherever you wish by cantilever. If you do it when the house is virtually finished, to get round planning permission, you can tell the planner ›the house was finished a week ago and we are now doing a permitted addition‹. You can add 140 sq.ft., or a good double bedroom in this way.« (See fig. p. 150.)

So with Lewisham Two, he was saying in 1984, »The houses are going to look quite different eventually, indeed in the not too distant future! Already in Lewisham One, we have three cases where houses have been added to; in one case, for example, a baby was born and another bedroom added (for £1200; by traditional means it would have been nearer £6000)....«

The advantages we have seen: the method is easily understood; it allows great flexibility in planning and in location. The lightweight construction is easily understood; it opens the method to use by all, regardless of strength, sex and age. It is economic, dry, pleasant to build and easily altered. There is virtually no waste; indeed, there is no »building site« as we know it.

On the other hand there are disadvantages. The construction is visually very dominant, particularly indoors. Traditional internal decoration, like wallpaper which does not come in »modular« width rolls, is terribly tedious to apply. The lightweight construction moves and squeaks; I have seen a needle jump off a gramophone record when someone walked heavily past in the corridor outside. Internal sound insulation is poor. Windows are intrinsically draughty, while the ventilated wall and roof voids inhibit insulation.[5]

Perhaps Segal at times pared down the sections too far; there are sagging roofs, there is at least one platform needing reinforcing with brick piers after a decade of use. Undoubtedly there are many »Segal« windows which don't slide after 4 or 5 years – and cladding panels which shouldn't, but do. But the remarkable advantages, aiding our regaining control over our own housing, give satisfactions too deep for these real disadvantages to displace.

But this itself raises a major issue which centres in that sensitive relationship of care, maintenance and inhabitation. Self builders, personally guided by Segal, »knew« their buildings, and gained confidence to improvise and adapt. But when, say, some years later a *new* owner moves in, there is a very different relationship with the building; different tolerances of, and sensitivity to what might be expected of the building.

»Technologically speaking,« said Segal, trying to define his achievement, »this means that you join high technology tools with very low technology applications. With that one, you can produce buildings as varied as anything you want.«

And to summarise his approach to building, it centres on systematisation without inventing a system; Segal used existing materials not components, transforming them by »seeing through« to a new application, rather than engaging in »second forming« their shape. Thus there is no »added value«.

»Standardisation in itself I have tried to do all my working life,« he said to me. »But in building it is only significant if *you* do not standardise but that you use standardised things.« The three-stage »building system« tradition – (1) starting with the material; (2) changing shape into components; (3) putting together on site – is reduced by Segal to two. »For me there is no intermediate stage; no components.«

Segal treats himself »like a consumer«; and rather than get involved in a supply industry of his own, he uses the already standardised bits available. In every situation he »sees through« to a radically original use of these as-found »ready-mades«.

Notes and Quotations

1. *The Architects' Journal*, 4 May 1988.
2. The term is medical; ›iatrogenic disease‹, is perhaps most clearly described in Ivan Illich *Medical Nemesis.*
3. I am considerably indebted to Jon Broome's »The Segal Method«, a special issue of *The Architects' Journal*, 5 November 1986 in this section.
4. Jon Broome, as 3.
5. I am grateful to Pete Sutton who, in his thoughtful study of one of the Lewisham houses (submitted towards the BSc architecture degree, PNL, March 1987), lists advantages and disadvantages.

Quotations from Walter Segal in this section from: Wills & Yeo interview 1984; 1971(B)2; 1976(C)2.

149

Segals »Pullover«, aus unbeschnittenen 50mm Holzwollplatten; dann eine Verkleidung, der »Regenmantel«, aus hochgepreßten Faserzement-Platten mit einer farbigen Emaillierung. »Früher konnte ich solche Qualitätsprodukte niemals benutzen, als ich noch an Kostenbeschränkungen gebunden war und deshalb billige und kurzlebige Produkte für die Außenhaut benutzen mußte.«

Typisch für die Innenhaut, die »Weste« sind unbeschnittene Gipsplatten, die zum Beispiel mit einer Holztafel für das Studio oder einer wasserfesten Beschichtung im Badezimmer versehen werden können. Spezielle Zwischenschichten, ein »Hemd« zwischen »Weste« und »Pullover« aus Polyurethanschaum zum Beispiel, können die Isolierung verbessern. Allerdings ist es immer noch eine Konstruktion, bei der es höhere Wärmeverluste gibt als in konventionell abgedichteten Bauten. Die thermischen Eigenschaften wurden seit dem Haus in Highgate verbessert, Jon Broome betont jedoch in seinem Artikel den Standpunkt der Verhältnismäßigkeit: »Segal wußte sehr gut, daß es einen Punkt gibt, an dem weiteres Ausfeilen die wirklichen Vorteile des Einfachen wieder zunichte machen würde.«

Das Fenster, seit Highgate raffinierter, ist ebenfalls nicht luftdicht, aber es kann auch als Doppelfenster eingebaut, oder es können einfach andere konventionelle Fenster verwendet werden. Das Segal-Fenster bleibt eine exemplarische Übung, wie mehr durch weniger zu erreichen ist: Die zwei Aluminiumwinkel, durch einen Dichtungsstreifen im Abstand gehalten, um das Schwitzwasser abfließen zu lassen, führen die Glasscheiben mit ihren polierten Kanten; es braucht nur ein leichtes Reiben mit einer Kerze, ehe man die Scheibe einsetzt.

Die Trennwände sind gleich wie die Außenwände konstruiert. Die üblichen Decken sind aus unbeschnittenen Gipsplatten, die lose auf Latten zwischen den Balken liegen. Alle Leitungen und Rohre können einfach in den leicht zugänglichen Hohlräumen des Fußbodens, der Wände und der Decken verlegt werden.

Segal liebte immer die Treppen und bewies große Findigkeit im Konstruieren. Er benutzte jede Gelegenheit, um kleine virtuose Pirouetten der Tischlerkunst in kleinsten Räumen zu zeigen. Innerhalb der Gebäudehülle können die Treppen, aufgelöst in einzelne Stufen, an der Decke oder an Auskragungen aufgehängt, oder sie können von Pfosten gestützt werden. Die Grundkonstruktion kann auch genutzt werden für Pergolas, Lauben, Veranden oder Sitzplätze. Segal sagte: »An einem Deckenbalken kann man alles aufhängen – Blumenfenster zum Beispiel, 600mm (ein Modul) breit, können das Haus, wo immer man will, durch eine Auskragung erweitern. Wenn man es tun möchte und das Haus schon praktisch fertig ist, kann man, um die Baugenehmigung zu umgehen, einfach behaupten ›Das Haus war schon vor einer Woche fertig, wir machen jetzt einen gestatteten Anbau‹. Auf diese Weise kann man bis zu 14m² anbauen, ein gutes Doppelschlafzimmer.«

»Bei Lewisham Zwei«, sagte Segal 1984, »werden die Häuser einmal ganz anders aussehen, und das in nicht allzu ferner Zukunft. Schon in Lewisham Eins haben wir drei Fälle solcher Erweiterungen. In einem Fall wurde ein Baby geboren, in einem andern ein Schlafzimmer angebaut (für £ 1200, bei traditioneller Bauweise wären es eher £ 6000 gewesen) ...«

Die Vorteile haben wir damit gesehen: Die Methode ist leicht zu verstehen. Sie erlaubt große Flexibilität in Grundriß und Situation. Auch die Leichtkonstruktion ist mühelos verständlich; sie kann von Menschen jeglichen Alters und Geschlechts gehandhabt werden. Sie ist wirtschaftlich, trocken, angenehm zu bauen und leicht zu ändern.

Doch gibt es auch Nachteile: Das Konstruktive wirkt sehr dominierend, besonders im Innern. Traditionelle Innendekorationen, wie Tapeten, die es nicht in der dem Modul entsprechenden Breiten gibt, sind sehr mühsam anzuwenden. Ich habe auch schon eine Nadel von der Grammophonplatte springen sehen, wenn jemand mit schweren Schritten draußen durch den Korridor ging. Die Leichtkonstruktion arbeitet und ächzt. Die Schallisolierung im Innern ist schlecht. Durch die Fenster zieht es, während die belüfteten Wand- und Deckenhohlräume die Isolierung erschweren.

Bay windows easily added.
Erkerfenster, auf einfache Weise hinzugefügt.

A pergola at the second Lewisham scheme.
Eine Pergola in Lewisham, zweite Bauetappe.

Twenty Years of Timber Buildings

Timber framing as a »natural« form for the small dwelling was always with Segal, and some years before developing his »method«, in 1957 he made two timber frame houses, each with Swiss connections.

The first was a ski house, high in the Swiss mountains almost a day's climb from road or train, built for himself and shared with another family. It typifies Segal's work. Simple yet appropriate, it stands with precision, 12m by 4m under the overhanging flat roof, on ten double stilts above the snow, its verandah floating within the frame of an almost immaterial handrail.

Unconnected to any service, it centres on a log-burning stove. Typically – like »La Casa Piccola« and »the little house« – it models in miniature a carefully arranged convivial home. The central space, with its thresholds to the outside in one direction gives to two cabins in the other direction, each with two upper and lower bunks. Beside each of the eight pillows is a tiny opening window; in the hub are two huge glass panes, costing as much as the rest of the house, which celebrate the magnificent view; »framing light so that you really notice it.«[1]

»It was rather like magnificent camping,« Moran Segal recalls. »We talked a lot, Walter and I, about comfort and striking the balance. He wasn't frugal or puritan, no; but simple. He had clear, simple requirements from life.«[2]

One day, showing photographs of the ski house to the Swiss who always sold him his annual holiday travel tickets, Segal found himself commissioned to design a »Swiss House«, in Twickenham, London.

This was an elegant balloon frame structure with hanging balconies, under a large monopitch roof and clad in boarding and brickwork below. The building inspector, to Segal's incomprehension, insisted on brick end walls. »It was the only part of the building he was interested in.«

With the »Segal method« timber houses, the planning is always refined, economical and thoughful, following the principles previously discussed. What was new, however, was the direct client, carpenter and architect relationship which made professional life so much more satisfying. Later, in looking back on this period, Segal noted that by the early 1960s, he »had become a much less amiable person than I am today. But now it seemed possible to escape from the treadmill of traditional practice.«

With each new client there was a pioneering venture in which Segal took part. But it was a journey without fear, based on Segal's practical confidence and experience. »I found that, unlike my traditional buildings, my clients now always knew how it would end. That they have kept very friendly relations with me was simply due to the fact that I did not turn them into nervous wrecks, that I did not make them fear all kinds of dangers, and they could see that things did perform as I had told them. In the end they were always delighted that it went quickly and that they were not surprised by any claims, additions and so on. I had complete control over expenditure.«

At one house, elegantly clad in red and white panels, as soon as it was complete, the client wanted two extra rooms; there were fruit trees he didn't want to cut down on the slope next to the house , and so the extension wrapped round them down the slope, with steps to join them and to a roof terrace above. It was very easy to make a new opening and reuse the wall unit in the extension. »This house taught me the degree of individualisation which a rigidly standardised method can achieve. There are only 18 elements in it.«

Segal was now involved in building a whole series of these lightweight, low-cost, spacious and light timber structures. For each one he would produce numerous plans with the clients, offering up to two dozen arrangements. Once he worked through 40 layouts with the clients – who added in another element: they wanted to build it themselves. So far Segal had run all the building works, organising craftsmen and subcontractors. He was unsure. »It was quite ambitions, they had never built before. I told him: ›You know, I just don't believe that you can put up the frame of this ambitions kind....

Zwanzig Jahre Holzbau

Holzbau, als eine natürliche Bauform für kleine Häuser, war Segal schon immer vertraut. Jahre bevor er seine »Methode« entwickelte, 1957, baute er zwei Häuser im Ständerbau, beide in Beziehung zur Schweiz. Das erste war ein Skihaus, hoch in den Schweizer Bergen, in der Nähe von Fideris. Er baute es für sich und eine andere Familie, und es ist typisch für Segal. Einfach, jedoch ausreichend, 12 x 4 m groß, unter einem weit auskragenden Flachdach, auf zehn Doppelstützen gestellt. Die Terrasse schwebt fast mit ihrem beinahe immateriellen Geländer.

Ohne Verbindung zu irgendwelchen Versorgungsleitungen ist das Skihaus um einen Holzofen gebaut. Wie bei der *Casa Piccola* und dem *Little House* ist es gewissermaßen eine Miniaturausgabe eines sorgfältig geplanten großen Hauses. Aus dem zentralen Raum geht es in einer Richtung direkt ins Freie, in der andern in zwei Kabinen, jede mit zwei Kajütenbetten. Neben den acht Kopfkissen gibt es acht kleine Fenster, jedes zum Öffnen. In den Zentralraum bringen zwei große Glasscheiben, die etwa soviel kosteten wie der Rest des Hauses, eine großartige Aussicht. »Wenn man das Licht einrahmt, nimmt man es erst wirklich wahr«, erzählte Moran Segal dem Autor 1988. »Es war eher wie ein großartiges Kampieren dort oben. Wir sprachen eine Menge, Walter und ich, über Bequemlichkeit und darüber, alles im Gleichgewicht zu halten. Er war nicht frugal oder puritanisch, nein, aber einfach. Er stellte klare, einfache Forderungen an das Leben.«

Im selben Jahr zeigte Segal eines Tages Fotos von diesem Skihaus einem Schweizer, der ihm alljährlich in London seine Fahrkarten verkaufte, und bekam auf der Stelle den Auftrag, solch ein »Schweizer Haus« in Twickenham zu bauen.

Dieses wurde eine elegante *balloon frame*-Konstruktion, mit aufgehängten Balkons unter einem großen Pultdach; darunter war es teils aus Backstein, teils verbrettert. Der Bauinspektor bestand auf Backstein für die Giebelwände. »Das war alles, was ihn interessierte.« Segal verstand das nicht.

Segal hat die Planung für die Holzhäuser immer weiter verfeinert. Er lernte in den direkten Beziehungen zu Bauherr und Zimmermann. Er sei, sagte er, »in den frühen 60er Jahren eine sehr viel weniger angenehme Person gewesen. Aber dann schien es möglich zu werden, der Tretmühle der traditionellen Praktiken zu entfliehen.«

»Anders als bei meinen traditionellen Bauten, wußten die Bauherren oft nicht, wie es enden würde. Daß wir trotzdem freundschaftlich miteinander auskamen, lag daran, daß ich keine nervösen Wracks aus ihnen machte, sie nicht mit allen möglichen Ängsten quälte, und daß sie sahen, daß die Dinge so funktionierten, wie ich ihnen gesagt hatte. Zum Schluß waren sie immer beglückt, daß es so schnell gegangen war und daß keine Nachforderungen kamen. Ich hatte immer alle Ausgaben unter Kontrolle.«

Bei einem Haus, das elegant mit weißen und roten Platten verkleidet war, wünschte der Bauherr, nachdem es gerade fertig geworden war, zwei weitere Räume. Am Hang neben dem Haus standen aber Obstbäume,

Ski house in Fideris, Switzerland, 1957.
Skihaus in Fideris, Schweiz, 1957.

Ski house in summer landscape.
Das Skihaus in der Sommerlandschaft.

Snow-covered landscape framed by the ski house great window.
Schneelandschaft, eingefaßt vom Rahmen des großen Fensters.

Tobler house, Twickenham, London, 1959. Garden front.
Tobler-Haus, Twickenham, London, 1959, Gartenseite.

I've got two carpenters who have done such things for me, and I would strongly suggest they should help you put up the frame and then you can go on.‹ And so it started, but in the evening of the first day, the client telephoned me: ›We are finding the carpenters too expensive, and we think we could do just as well; so, if you have no objection, we'd like to send them away.‹

»I can't say very much! ›Well‹, I said, ›this is your house, so you do as you think right‹. So the carpenters were sent away, and the clients built it. Which, if you think that they were two schoolteachers never having used tools in their lives, they did rather undertake a bold affair. But with their enthusiasm and motiovation, there was no trouble and no difficulty and it succeeded quite astonishingly.« That was the first »self-build«. Soon another client couple wanted to build theirs; Segal thought, »All right, perhaps they can really do it. Let us see, they may organise it.« And he asked them to make their own plan layout. Segal claimed he simply tidied up the plumbing. That worked too.

Then an Irishman wanted to build his three-bedroom house with a carpenter friend during his three weeks' holiday. A design having been agreed upon and materials organised, Segal went over to set it out with him. A day and a half later, with the frame up, Segal left saying »You now have 19 1/2 days in which to complete the building.« He did it. »I learned from this,« recalled Segal, »what a man using unsophisticated technology can produce in a very short time. He invited me to come over and stay in the house. He had made practically no mistakes. This house greatly encouraged my confidence in the fact that people can help themselves.«

A large house was built in Bedfordshire by a couple who wanted a pitched roof. Its design intrigued Segal, who felt that the flat roof provided optimal performance but by now he was happy to go along with any client preference like this. He was even more intrigued to observe the »young, diminutive Welsh wife who did most of the supporting structure with her own hands – I believe there were some 9,000 nails which she had to

die er nicht fällen mochte; so wurde die Erweiterung einfach um diese herum den Hang hinab gebaut. Es war ganz einfach, eine neue Öffnung zu machen und das Wandelement für den Anbau wieder zu verwenden. »Dieses Haus lehrte mich den hohen Grad an Individualität, den man mit einer streng genormten Methode mit nur 18 Elementen erreichen kann.«

Segal baute eine ganze Serie dieser leichten, preiswerten, geräumigen und lichten Holzhäuser. Für jedes einzelne produzierte er zusammen mit dem Klienten zahlreiche Pläne, oft bis zu zwei Dutzend Varianten. Einmal arbeitete er sich mit den Kunden durch 40 verschiedene Grundrisse. Schließlich wollte der Kunde das Haus selber bauen. Segal war unsicher. »Es war ein ehrgeiziges Unterfangen; er hatte niemals zuvor etwas gebaut. Ich sagte ihm: Sie wissen, ich kann mir einfach nicht vorstellen, daß Sie solche Rahmen selber aufstellen können ... Ich habe zwei Zimmerleute, die so etwas schon für mich gemacht haben, und ich würde Ihnen sehr empfehlen, daß Sie sich helfen lassen, die Rahmen aufzustellen, und dann können Sie weitermachen. Aber am Abend des ersten Arbeitstages telefonierte der Kunde: Wir finden die Zimmerleute zu teuer, und wir glauben, wir können es ebenso gut wie sie. Wenn Sie also nichts dagegen haben, möchten wir sie gerne wegschicken. Ich konnte nicht viel sagen. Gut, sagte ich, es ist Euer Haus, tun Sie was Sie für richtig halten. Also wurden die Zimmerleute fortgeschickt, und die Hausherrn bauten selber. Wenn man denkt, daß beide Lehrer waren, die niemals in ihrem Leben ein Werkzeug benutzt hatten, war das sicher ein kühnes Vorhaben. Aber mit ihrer Begeisterung und ihrer Motivation gab es keine Probleme und keine Schwierigkeiten, und es ist ihnen erstaunlich gut gelungen.«

Das war der erste Selbstbau. Bald darauf kam ein anderes Ehepaar, das sich ein Haus bauen wollte. Segal dachte »Gut, vielleicht können sie es wirklich, laßt uns sehen, wie sie sich organisieren.« Und er forderte sie auf, ihren eigenen Grundriß zu entwerfen. Segal behauptete später, er hätte nur die Installationen etwas entwirren müssen.

Dann wollte ein Ire sein Vierzimmerhaus zusammen mit einem befreundeten Zimmermann während seiner drei Ferienwochen bauen. Nachdem man sich auf den Grundriß geeinigt und das Material organisiert hatte, fuhr Segal nach Irland, um das Haus abzustek-

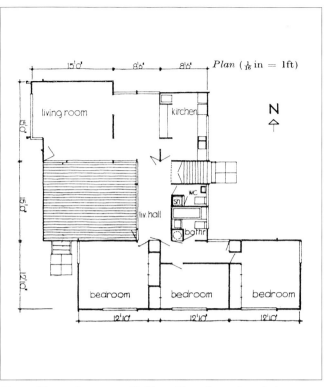

Plan of large house at Yelling. Note how Segal develops the theme of separating the communal living house from the dormitory house with an open room (variously called court, patio, balcony, deck).
Grundriß eines großen Hauses in Yelling. Segal trennt den gemeinsamen Wohnteil von den Schlafräumen mit einem offenen Raum ab, den er wechselweise »Hof«, »Patio«, »Balkon« oder »Deck« nennt.

ken. Eineinhalb Tage später standen die Rahmen. Segal verließ den Bauherrn und sagte: »Jetzt habt Ihr 19 und einen halben Tag, um das Gebäude fertigzustellen.« Sie schafften es. »Ich habe hieraus gelernt«, erinnerte sich Segal, »was ein Mensch mit einer nicht zu raffinierten Technologie in ganz kurzer Zeit fertigbringen kann. Der Ire lud mich später ein, ihn zu besuchen. Er hatte praktisch keine Fehler gemacht. Dieses Haus stärkte wirklich mein Vertrauen in die Fähigkeit der Menschen, sich selbst zu helfen.«

Ein großes Haus wurde von einem Paar in Bedfordshire gebaut. Sie wollten ein geneigtes Dach. Der Entwurf reizte Segal, der zwar glaubte, daß das Flachdach die optimale Lösung sei, aber jetzt doch gerne auf

Yelling house. The living room seen from the deck.
Haus in Yelling. Der Wohraum vom »Deck« aus.

put in.« Another couple made their frames and or-ganised friends to help them raise them. »I took another self-builder, Ken Atkins, with me,« said Segal. »The friends hadn't turned up, and the clients were getting anxious.... So Ken went to the pub and came back with five half drunk young men to whom he's promised a fiver if they'd help. He organised them, and in two and a half hours the whole thing was up. The clients were delighted!« One has been built in Perth, Australia, based on a hour and a half telephone call in 1973. »Stu-dents at Western Australia University managed to get hold of a telephone, apparently,« said Segal most amused. »Now *that* is a splendid house!«

One couple in a »Segal method« house had a party. »All the wife's family from the Auvergne were com-ing. They could have hired a room, of course, but in-stead took out all the partitions in their house, stood them against the wall and had the whole space free. Filled with tables and so forth, the whole congregation could celebrate the family day at home – and afterwards it was put back in two hours. This simmered in my mind for a number of years,« added Segal. The son of that same family was getting married; the client rang Segal: »Can we take down our house and move it to the other end of our plot and rebuild it? We *like* our own house, but prefer the other end of the plot; and then there'd be room, please, if you'd design us a new house for our son.« Segal was enthusiastic, but sadly the project didn't materialise.

Another family bought a second-hand freezer; how could they get it into the house? They might have

Yelling house. Inside the living room.
Haus in Yelling. Wohnraum mit Blick gegen »Deck«.

alle derartigen Wünsche seiner Kunden einging. Mehr noch fesselte es ihn, »die junge, winzig kleine Frau aus Wales zu beobachten, die fast die ganze Tragkonstruktion allein zusammensetzte – ich glaube, mehr als 9000 Nägel hatte sie einzuschlagen.«

Ein anderes Paar hatte die Rahmen zusammengesetzt und Freunde organisiert, sie aufzurichten. »Ich ging zusammen mit einem anderen Selbstbauer, Ken Atkins, hin«, sagte Segal. »Die Freunde waren nicht gekommen, und die Bauherrn wurden unruhig. Da ging Ken ins nächste Gasthaus und kam mit fünf halb betrunkenen jungen Männern zurück, denen er einen Fünfer versprochen hatte, wenn sie helfen würden. Er organisierte sie, und in zweieinhalb Stunden war alles aufgerichtet. Die Bauherren waren begeistert.«

»Ein Haus ist in Perth, in Australien, gebaut worden, 1973; es basiert auf einem anderthalbstündigen Telefongespräch. Studenten der Universität von West-Australien hatten ein Gratis-Telefon in die Hand bekommen«, sagte Segal, außerordentlich amüsiert. »Jetzt ist das ein großartiges Haus.«

Ein Ehepaar in einem Segal-Haus hatte eine Party. »Die ganze Familie der Frau aus der Auvergne kam. Man hätte natürlich einen Saal mieten können; aber stattdessen nahmen sie alle Trennwände weg, lehnten sie an die Wand und hatten den ganzen Raum frei. Sie stellten Tische und alles andere auf, und die ganze Gesellschaft konnte das Familienfest zu Hause feiern. Hinterher wurde alles in zwei Stunden wieder eingebaut.«

Als der Sohn derselben Familie heiratete, telefo-

Wembley playroom, outside view.
Spielhaus in Wembley, Außenansicht.

been able, with extreme effort, to manoeuver and manipulate it up the steps and round the angles to the kitchen. Instead they took out three panels, lifted it in and put the panels back. The operation took about 40 minutes.

By the early 1970s, Segal was buoyant. »I have now gained sufficient confidence in the method of building,« he said in 1971, »to go beyond the building of individual houses, and tackle larger projects such as housing and schools. Though I do not think it will be possible for such simple technologies to reverse the whole trend of housing, they can make a very large contribution to it. We need to bring back into mass housing, building of smaller scale and new thinking about layouts.«

Segal became engrossed in producing numerous layouts of both single and two-storey, high-density housing built by his method. At last the ideas could move on again from the individual house.

Notes and Quotations

1. John Segal adds: »Though I doubt very much that they cost as much as the rest of the house, I was always most impressed by the feat of transporting these enormous double-glazed units up that mountainside on a horse-drawn sledge without breakage.« Letter to the author, 27 August 1988.
2. Moran Segal, in conversation with the author, April 1988.

Quotations from Walter Segal in this section from: 1977(B)2; 1983(B); 1971(B)2; Wills & Yeo interview 1984.

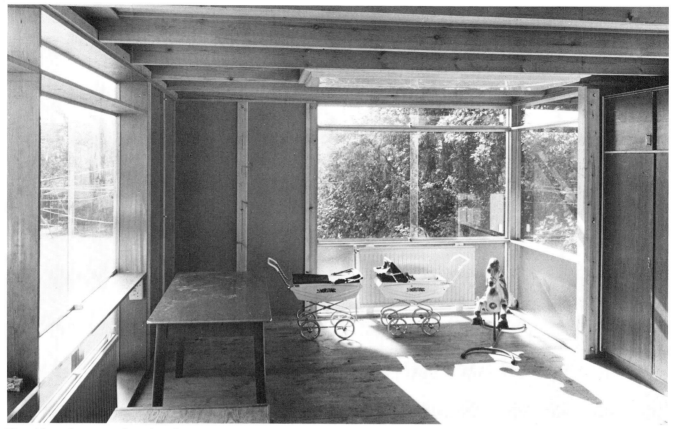

Wembley playroom, interior view.
Spielhaus in Wembley, Innenansicht.

nierte der Vater an Segal: »Können wir unser Haus ab-
montieren und am anderen Ende des Grundstücks
wieder aufbauen? Wir lieben unser Haus, würden es
aber lieber dort haben. Und dann wäre genügend Platz
da für ein zweites Haus. Wenn Sie, bitte, ein neues Haus
für unseren Sohn planen würden.« Segal war begeistert; aber leider wurde nicht daraus.

Eine andere Familie kaufte einen gebrauchten
Kühlschrank. Aber wie konnte man ihn ins Haus bekommen? Sie hätten ihn mit äußerster Anstrengung irgendwie die Treppe hinauf und um die Ecken in die Küche
schaffen können. Stattdessen nahmen sie drei Paneele
heraus, hoben den Schrank von außen in die Küche und
setzten die Paneele wieder ein. Die ganze Operation
dauerte 40 Minuten.

Anfang der Siebziger Jahre war Segal in gehobe-
ner Stimmung. »Ich habe jetzt genug Vertrauen in
diese Baumethode gewonnen«, sagte er 1971, »um
über das Bauen von Einzelhäusern hinauszugehen,
und größere Projekte in Angriff zu nehmen, wie Siedlungen und Schulen. Obwohl ich nicht glaube, daß
es möglich sein wird, mit einer so einfachen Technologie den gesamten Trend im Wohnungsbau umzu-
kehren, kann sie doch einen erheblichen Beitrag dazu
leisten. Wir müssen nun den Wohnungsbau für die
Massen mit Bauten in kleinerem Maßstab und mit
neuen Überlegungen in Angriff nehmen.« Endlich
konnte er seine Ideen weiterentwickeln, über das individuelle Haus hinaus.

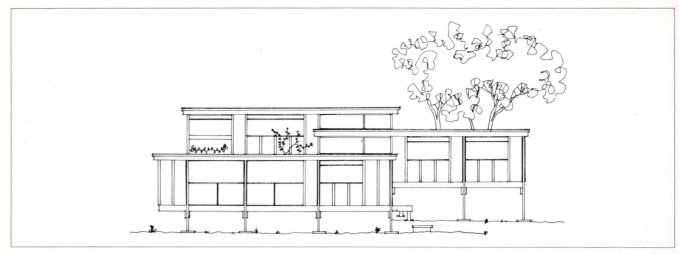

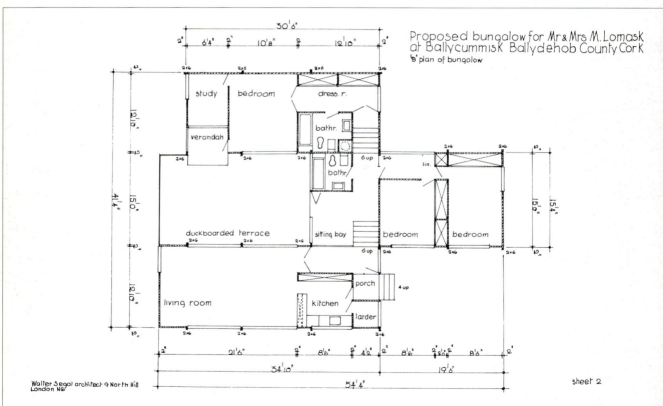

Proposed bungalow for Mr & Mrs M. Lomask
at Ballycummisk Ballydehob County Cork
'b' plan of bungalow

study   bedroom   dress. r.

verandah

bathr.

bathr.

duckboarded terrace   sitting bay   bedroom   bedroom

lin.

living room   kitchen   porch

larder

Walter Segal architect 9 North Hill
London N6

sheet 2

House at Ballycummisk, Ireland. Elevation and plan showing the terrace enclosed on the level between sleeping house and living house.
Das Haus in Ballycummisk, Irland. Ansicht und Grundriß. Der Grundriß zeigt die ins Haus eingeschobene Terrasse, die den Schlaf- vom Wohnteil trennt.

Birch house, Mill Hill, North London, under construction; the first Segal two-storey frame. (The architect appears in both interior and exterior view.)
Das Haus Birch in Mill Hill, Nord-London, während der Bauzeit; Segals erstes zweigeschossiges Fachwerkhaus. (Der Architekt erscheint auf beiden Bildern.)

Self-Build from the Housing Authority

Land

Of course, undermining all the effort to produce homes as economically as possible, was the unspoken factor of land. Segal had taken part in the post-war thrust which very nearly made land nationalisation in Britain a reality. This would have done more than any other arm of national policy to secure the future of a more equitable nation, as well as one where good housing was part of the everyday equipment of everyone.[1]

In 1945, in *Planning and Transport*, Segal wrote »only a National Plan [based on public ownership of land] will cope efficiently with the intricacies of land utilization.« If it had very nearly been achieved around 1950, it was quickly unravelled between 1951 and 1953; so that within two decades it had become a lost dream no longer on anyone's agenda; and land costs were spiraling upwards.

By 1971, »in Britain subsidised housing is nowadays in dire jeopardy because of the impossible situation that has been permitted to develop on the land market. To redress the balance some sort of *national sacrifice* will have to be made to get hold of, and to control, land nationally.« [Segal's italics]. This issue threatened the basis of Segal's efforts. He often expressed horror when finding a house which he had built with immense economy, to be for sale thirty years later for 100 times its initial cost.[2] And this spiral is firmly centred on the tight control on land, and its artificial scarcity.

In 1971 Segal said:»It sounds absurd that one should try to search for methods of building cheaply when we have the impossible problem of accelerating land cost and do nothing about it. My buildings and methods will in the end be strangled by this problem unless something is done to overcome it.« In 1945 Segal had written that land accounted for 5% of the cost of housing, roads and services 15%, and the actual building 80%; but by the early 1960s land was costing nearly 40% of the total. Meanwhile those homes he built in the early 1960s were being resold only seven years later at 400% of what they had cost.

Segal was concerned that we free housing from »its parasitical constraints, and augment the land available for building. Local authorities possess a fairly considerable amount of land which should pass into national possession; legally this would not be a fantastic problem and can be achieved within existing frameworks. These ›national land banks‹ will grow by buying land, not appropriating it.«

Thus Segal began to sketch a possible context for his own work: »Such land could then be made available to groups and individual users on the basis of leases which would be related to the expected life of the buildings, though with preferred methods of construction. Such leases could be shorter than leases in use at the present, perhaps not longer than the span of one generation. The bulk of the available subsidies should, with as few strings as possible, permit applicant users to construct *vetted* housing outside the stereotype framework in which authorities operate.«

He was well aware that, in Martin Pawley's words, »his system of house design and construction – but for the insane economics of the housing market today – could see every homeless family in Britain housed in a dwelling of their own design and construction by the turn of the century.«[3]

In 1974 Segal talked at the Dweller Control Housing Group which John F C Turner had set up at the Architectural Association[4]. He aired his hope to sell self-build to housing authorities; he foresaw groups of 10 to 20 houses, short-lease (stressing public ownership of land), and costing between a quarter and a third of current public authority dwellings. Colin Ward, radical writer on housing and self-help, reported Segal's ideas to his friend Brian Richardson, deputy architect of Lewisham, and also suggested the possibility of Lewisham to Segal.[5]

Doomed projects

With that seed sown, Segal kept working with immense energy through the mid 1970s to get any project off the ground, but time and again was thwarted. His complete

Selber-bauen = Sich-selber-bauen

Land

Jedoch gab es ein Hindernis, das den Bemühungen, Wohnungen so wirtschaftlich wie möglich zu bauen, den Boden entziehen konnte: das Problem des Baulandes. Segal hatte nach dem Kriege die Vorstöße unterstützt, die fast die Nationalisierung des Bodens hätten Wirklichkeit werden lassen. Das hätte mehr als irgend etwas anderes dazu beigetragen, eine gerechtere Nation einzurichten, ein Land, in dem eine gute Wohnung für jeden etwas Alltägliches hätte werden können. (Siehe *Homes for the People*, S. 150 und 173, Redaktion Andrew Boyd und Colin Penn, *Association of Building Technicians*, pub. Paul Elek, 1945.)

Das Buch *Homes for the People*, an dem Segal 1945 mitarbeitete, rief nach einem nationalen Plan, der auf dem öffentlichen Eigentum des Bodens beruhen sollte. Nachdem das Ziel um 1950 fast erreicht worden wäre, war es um 1970 ein ausgeträumter Traum. Die Landkosten schraubten sich mehr und mehr in die Höhe.

Das bedrohte die Basis von Segals Bemühungen. Er zeigte sich oft entsetzt, wenn ein Haus, das er mit äußerster Sparsamkeit gebaut hatte, 30 Jahre später für das 100-fache der ursprünglichen Kosten zum Verkauf kam. Sein eigenes Haus in St Annes' Close war ein typisches Beispiel: für £ 2500 gebaut, und, nachdem das Baurecht nach 30 Jahren abgelaufen war, war es £ 250 000 wert.

1971 sagte Segal: »Es klingt absurd, daß man nach Methoden suchen soll, billiger zu bauen, solange wir dieses unmögliche Problem der steigenden Landkosten nicht beseitigen können. Meine Bauten und Methoden werden durch dieses Problem abgewürgt.« Er erklärte, daß, als er ein junger Architekt war, die Landkosten etwa 5%, die Kosten für Straßen und Leitungen 15% und die Kosten für das eigentliche Gebäude 80% betragen hätten; aber in den frühen Sechziger Jahren war der Anteil der Landkosten schon 40%.

Es war Segals Anliegen, »das Wohnen von parasitären Einschränkungen zu befreien, für das Bauen mehr Land zur Verfügung zu stellen. Die lokalen Behörden verfügen über eine erhebliche Menge an Land, das in nationalen Besitz übergehen sollte. Rein legal wäre das kein so ungeheures Problem und innerhalb des bestehenden Rahmens möglich. Diese nationale Landreserve sollte durch Landkäufe, nicht Enteignungen wachsen.«

So begann Segal ein denkbares Umfeld für seine Arbeit zu skizzieren: »Solches Land sollte dann Gruppen oder individuellen Nutzern auf der Basis eines Baurechts, etwa entsprechend der Lebensdauer des Gebäudes, zur Verfügung gestellt werden. Solche Verträge könnten kurzfristiger sein als heute, vielleicht nicht länger als die Spanne einer Generation. Der Hauptteil der zur Verfügung stehenden Subventionen sollte, mit so wenig Einschränkungen wie möglich, den Bewerbern erlauben, ihre Wohnungen nach gültigen Methoden auch außerhalb des stereotypen Rahmens zu bauen, in dem Behörden operieren.«

Er war sich der Tatsache wohl bewußt, daß sein System des Entwurfs und der Konstruktion es ermöglichen könnte, bis zur Jahrhundertwende jede wohnungssuchende Familie in einem selbst entworfenen und selbst gebauten Heim zu haben – wären die ökonomischen Verhältnisse auf dem Wohnungsmarkt nicht so unsinnig wie sie heute sind. (*The Guardian*, 2. 6. 1988)

1974 sprach Segal vor der *Dweller Control Housing Group*, die John F.C. Turner (bekannt durch Selbstbau-Projekte in Lateinamerika und England, Mitarbeiter von »Freedom to Build«) bei der *Architectural Association* ins Leben gerufen hatte. Er äußerte die Hoffnung, den Selbstbau einmal den Wohnbaubehörden verkaufen zu können. Er dachte an Gruppen von 10 bis 20 Häusern auf staatlichem Land mit relativ kurzem Baurecht, die Wohneinheit zu etwa einem Viertel bis einem Drittel der Kosten im heutigen sozialen Wohnbau. Colin Ward, ein radikaler Schreiber zum Thema Wohnung und Selbsthilfe, sprach über Segals Ideen mit seinem Freund Brian Richardson, stellvertretendem Stadtarchitekten von Lewisham, und schrieb darüber an Segal: »Ich sagte ihm, daß nach meiner Ansicht die Leute in

The paired houses at Longton Avenue, Lewisham. Garden view of these two self-built houses.
Das Doppelhaus an der Longton Avenue in Lewisham. Gartenseite dieser beiden Selbstbau-Häuser.

frame of reference, despite its utterly clear internal logic, was radically alien to established practice; and it demanded an equally unprejudiced response. The Lewisham councillor who helped it finally to bear fruit there talked of Segal's »suicidal enthusiasm.«[6]

This may contain a germ of truth, but the frustration must have been immense when, after slow, careful and gentle development, project after project foundered. In 1975, three very different but sizable designs were simmering.

At Sussex University, a large anonymous donation launched the project of a student social centre, the donation being conditional on its being self-built, and the donor recommended an investigation of the »Segal method«. However, after Segal had been approached, the student body vetoed the self-build as it would be depriving building workers of jobs. The major donor withdrew, Segal continued with the design but the university demanded he go to competitive tender, making nonsense of the whole »Segal method«. This caused interminable delays after which another sponsor withdrew. Even after all that, in May 1976, the project was about to start on site when it finally ran out of breath and died.

In 1975, Segal had talked at a summer school for the Centre for Alternatives in Urban Development in Swindon, and this led to the »Segal method« becoming adopted by the Tothill Self-build Housing Association there. It took a patient year to reach the stage of appointing him architect. Then, with land agreed upon, the plans for a cul-de-sac of 11 single-storey houses

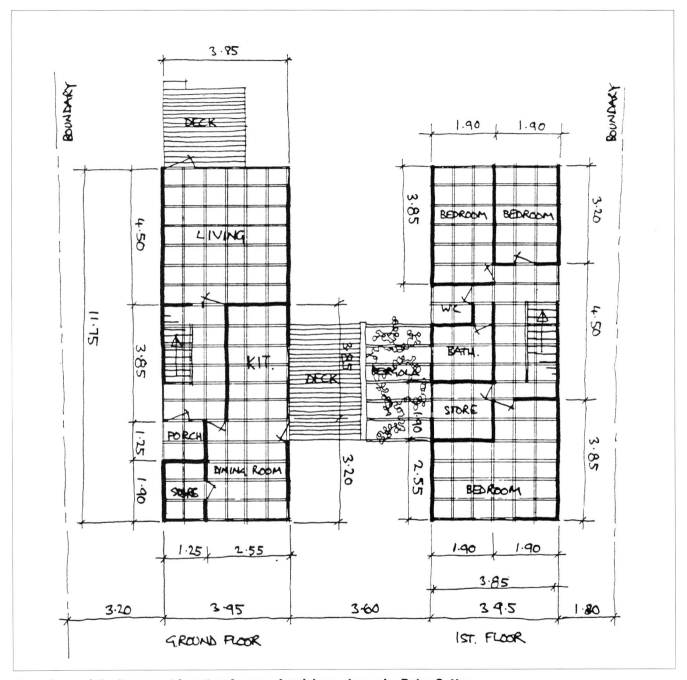

Plan of one of the houses at Longton Avenue, Lewisham, drawn by Peter Sutton.
Grundriß eines der beiden Häuser an der Longton Avenue, Lewisham, gezeichnet von Peter Sutton.

were quickly finalised with the self-build group, and presented, with the quantity surveyor's financial documentation, to the local authority.[7]

Spacious three-bedroom houses would cost £9500 each (including land), take six months maximum to build, and have a market value of around £15000. At just this moment, local builders, unable to sell their »first time homes«, reduced prices to around £9000, offered loan assistance, and all the about-to-be-married couples in the Self Build Association (who were a majority) left to buy tiny brick houses. Again, this Segal project hobbled slowly to a standstill.

Also in 1975, Segal presented a scheme of 10 two-storey timber-frame houses for a co-operative housing association, to be built in the outer London borough of Bromley. Here the dense even-tempered »carpet« layout and the two-storey flat-roof construction would be breaking new ground. It took three further years of increasingly acrimonious negotiation with local official-dom for this finally to fizzle out. During Segal's battle with the local planning officers and the government's Department of the Environment, the latter suggested that in the interim, perhaps »something more conventional be designed.«[8] (See figs. p. 170.)

In 1976 Segal designed a self-build scheme of 10 houses in Acrington which never materialised. For the Society for Co-operative Dwellings he produced a dense carpet of single- and two-storey patio houses in south London which was brought to a halt by bureaucracy, leading the impatient client to choose a »more conventional« solution.

Lewisham

The Lewisham seed, however, just survived. One of the borough's young architects, Jon Broome (who had also been attending »dweller control« meetings at the AA), with whom Richardson discussed Segal, had already been approached by a residents' group needing a community/sports pavilion. Richardson and Broome thought here might be an ideal way to introduce »Segal method« building gently into Lewisham.

But Walter Segal's letter to Richardson in October 1974 firmly replied: »I'm not keen to do it. I regard the technology as secondary. My interest is in self-help house-building for a Friendly Society on leasehold land.... I want to use my building method in connection with making a community.«

When eventually Ward arranged for Segal and Richardson to meet at a party, in July 1976, »the two individualists hit it off,«[9] and things started moving. Leading councillors were wooed, they called for a committee report, and Richardson's report recommending the Segal way came to the council. Segal addressed a Lewisham meeting; councillors and housing officials visited Segal's first self-build house; it was a fine day, Segal's clients enthused and plied the party with cups of tea. »This visit was the turning point,« said Richardson.[10] His report was approved by the Housing Committee by a single vote on 9 March 1976.

Lewisham saw three advantages to themselves: it would use otherwise dead land, it would be cheap housing, and they would have no maintenance responsibility. In the end, it was not to prove much cheaper than the »housing cost yardstick«, once all the authorities had been satisfied and the self-build labour was costed in. The crucial financial advantage to the authority was that the »Segal method« inevitably gave a fixed price. There would be no claims, no hidden additional costs; it would be built on budget. (This was almost unknown in local housing authorities, and *that* is how the cost per dwelling came to about 60% to 75% of average comparable London housing costs.)

In July 1976, a public meeting was announced at which 168 of those on the council housing list turned up. Equity sharing was agreed upon and a Self-build Association set up; at a second meeting in December 1976, a ballot selected the initial group for the 14 sites which Lewisham had offered.

Thereafter excitement within the group grew. Regular meetings began, with seminars and workshops where Segal explained his building system. Tradespeople were hired to teach elementary plumbing and electrics, and they visited other Segal houses.

The »unconventional« method, however, nearly stopped this project like others in the bureaucratic mire. There was no mechanism for processing the funding, the building inspectorate were unfamiliar with the construction, the authority couldn't work out how to order and pay for material; crucially, the government Department of the Environment which had offered immediate support through its regional architect, let this architect retire soon thereafter, to be

Camden Labour-Idealisten seien mit einer Mission, nämlich jedem eine Wohnung zu verschaffen.«

Untergegangene Projekte

Nachdem die Saat einmal gesät war, arbeitete Segal mit enormer Energie weiter, um ein solches Projekt in Gang zu bringen; aber immer wieder wurde es verzögert. Das ganze Gerüst seiner Ideen, trotz aller Logik, war den etablierten Praktiken diametral entgegengesetzt. Der Stadtrat von Lewisham sprach von Segals »selbstmörderischer Begeisterung«. (Nicholas Turner, *Learning the Lewisham Way*, The Architects' Journal, 18. 5. 88/S. 87)

Segal hatte zur selben Zeit die Frustration zu verarbeiten, daß drei langsam, sorgfältig und behutsam entwickelte Projekte scheiterten.

An der Sussex Universität hatte eine große anonyme Stiftung das Projekt eines sozialen Zentrums an die Hand genommen, unter der Bedingung, daß es im Selbstbau erstellt werden müsse. Der Donator hatte angeregt, die Segal-Methode zu studieren. Nachdem man sich mit Segal schon in Verbindung gesetzt hatte, legte die Studentenschaft ein Veto gegen den Selbstbau ein, da man damit Bauarbeiter um ihren Job gebracht hätte. Darauf zog sich der Spender zurück. Segal bearbeitete den Entwurf weiter; aber die Universität verlangte eine offene Ausschreibung und führte damit die Segal-Methode ad absurdum. Die Auseinandersetzung führte zu unendlichen Verzögerungen, bis ein weiterer Sponsor ausschied. Trotz allem hätte man im Mai 1976 mit dem Bau beginnen können; aber dann ging das Geld aus, und das Projekt war endgültig erledigt.

1975 hatte Segal vor der Sommerschule des *Centre for Alternatives in Urban Development* in Swindon gesprochen. Die Folge war, daß sich die *Self-build Housing Association* in Tothill für die Segal-Methode interessierte. Es brauchte ein ganzes Jahr und viel Geduld, ehe man so weit war, einen Architekten zu bestimmen. Dann, nachdem auch das Land zur Verfügung stand, wurden die Pläne für 11 eingeschossige Häuser an einer Sackgasse mit der Selbstbauer-Gruppe fertiggestellt und zusammen mit den Kostenschätzungen der Baubehörde unterbreitet (Peter Stead: *Self-Built Housing Groups and Co-operatives*, 1979): Geräumige Häuser mit vier Schlafzimmern sollten je £ 955, inklusiv Land, kosten, in maximal sechs Monaten gebaut sein und einen Marktwert von rund £

15000 darstellen. Und gerade in dem Augenblick reduzierten lokale Unternehmer die Preise für ihre »Häuser zum Anfangen« auf rund £ 9000 und boten Finanzierungshilfen. Die meisten der Paare, die gerade vor der Hochzeit standen – und das war die Mehrzahl – kauften die winzigen Backsteinhäuser der Bauunternehmer. Damit kam auch dieses Segal-Projekt zu seinem Ende.

Ebenfalls 1975 stellte Segal ein Projekt für 10 zweistöckige Häuser in Holzbauweise vor. Das Projekt hätte in einer äußeren Londoner Vorortsgemeinde, in Bromley, von einer kooperativen Wohnbaugesellschaft realisiert werden sollen. Hier wäre die dichte Flachdach-Teppichsiedlung etwas ganz Neues gewesen. Nach drei Jahren zunehmend schärfer werdenden Aus-

The first Segal self-built house. Ken Atkins on the roof with Jon Broome and on the ladder Walter Segal.
Das erste Segal-Selbstbauhaus. Ken Atkins auf dem Dach mit Jon Broome und auf der Leiter Walter Segal.

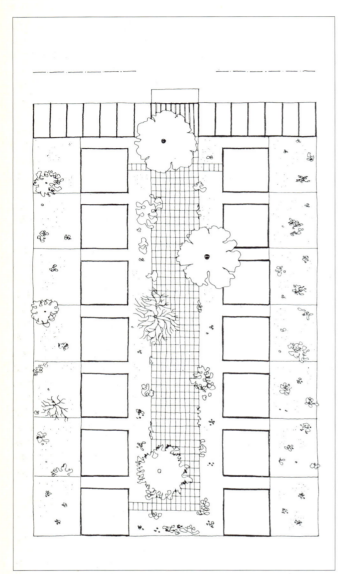

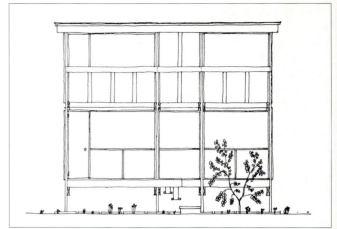

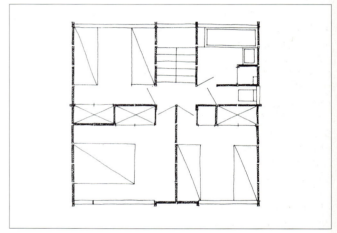

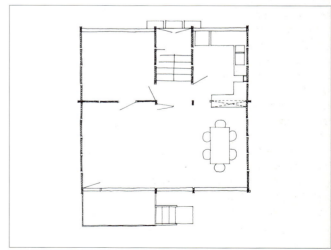

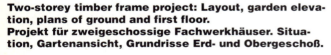

Two-storey timber frame project: Layout, garden elevation, plans of ground and first floor.
Projekt für zweigeschossige Fachwerkhäuser. Situation, Gartenansicht, Grundrisse Erd- und Obergeschoß.

einandersetzungen mit der lokalen Beamtenschaft war es auch damit zu Ende. Während Segals Kampf mit der lokalen Baubehörde suggerierte das Umwelt-Departement der Regierung, daß man während dieser Zeit doch vielleicht etwas Konventionelleres planen könne (John McKean: *The Anarchy of Planning*, Building Design 17. 3. 78, S. 14).

Lewisham

Die Lewisham-Saat jedoch überlebte. Einer der jungen Stadtarchitekten, Jon Broome, stand bereits in Verbindung mit einer Bürgergruppe, die ein neues Gemeinschafts- und Sportgebäude wünschte. Richardson und Broome dachten, daß das eine ideale Gelegenheit wäre, die Segal-Methode behutsam in Lewisham einzuführen.

Aber in seinem Brief an Richardson vom Oktober 1974 antwortete Segal bestimmt: »Ich möchte das nicht tun. Ich betrachte die Technologie als zweitrangig. Mein Interesse gilt dem Wohnungsbau durch Selbsthilfe für eine freundliche Gesellschaft auf Land im Baurecht . (...) Ich möchte meine Baumethode im Zusammenhang mit dem Ziel einer neuen Gemeinschaft einsetzen.«

Als Ward dann arrangierte, daß Segal und Richardson an einer Party im Juli 1976 zusammentrafen, »fanden die beiden Individualisten sofort zueinander«, (aus einem Brief von Richardson an Ward, Sommer 1976, veröffentlicht in Richardson: *Letters from a Papermakers' Husband*, Matrix, Wittington Press), und die Dinge kamen in Bewegung. Bei wichtigen Stadträten wurde angeklopft; diese verlangten einen Bericht des Bauausschusses. Der Bericht empfahl die Segal-Methode. Segal sprach vor einer Versammlung in Lewisham. Stadträte und Beamte der Wohnungsbehörde besuchten Segals erstes Selbstbauhaus; es war ein schöner Tag, Segals Bauherren, begeistert, versorgten die Herren mit Tee. »Dieser Besuch war der Wendepunkt«, sagte Richardson (in einer Unterhaltung mit dem Autor am 28. 2. 1988). Sein Bericht wurde am 9. 3. 1976 einstimmig vom Wohnungs-Ausschuß angenommen.

Die Leute in Lewisham sahen für sich drei Vorteile: Es würde bis dahin brachliegendes Land genutzt werden, es würde billige Wohnungen geben, die Stadt würde nicht für den Unterhalt verantwortlich sein. Am Schluß, nachdem man alle Auflagen erfüllt und die Eigenleistungen eingerechnet hatte, wurden die Bauten nicht viel billiger als nach dem *Housing Cost Yardstick*

der Behörde. Ein wesentlicher Vorteil für die Behörde aber war, daß es bei Segal einen fixen Preis gab. Es würde keine Nachforderungen, keine versteckten Mehrkosten geben; es würde alles im Rahmen des Budgets sein. So etwas kannten die Behörden kaum. Und damit reduzierten sich auch die Kosten pro Wohnung auf 60–75% des vergleichbaren Londoner Durchschnitts.

Im Juli 1976 wurde eine öffentliche Versammlung angesetzt, und 168 Personen von der Warteliste der Wohnungsbehörde kamen. Man gründete eine Selbstbauorganisation. Bei einem zweiten Treffen im Dezember 1976 wurde durch das Los die erste Baugruppe für die zur Verfügung stehenden 14 Grundstücke ausgewählt.

Danach wuchs die Aufregung in der Gruppe. Es begannen regelmäßige Treffen mit Seminaren und Workshops, bei denen Segal sein System erklärte. Handwerker wurden angestellt, um Grundkenntnisse über Elektrizität und Sanitärarbeiten zu vermitteln. Andere Segal-Häuser wurden besucht. Die unkonventionelle Methode jedoch ließ auch dieses Projekt fast wieder im Sumpf der Bürokratie versinken. Es gab kein Vorbild für die Finanzierung. Das Bauinspektorat war mit der Konstruktion nicht vertraut. Die Behörden wußten nicht, wie sie das Material bestellen und bezahlen sollten. Besonders kritisch: Das Departement für Umwelt, das zunächst volle Unterstützung durch eine regionale Mitarbeiterin angeboten hatte, zog diese zurück und ersetzte sie durch einen entscheidungsunfähigen Beamten, der sich auf einen engstirnigen Techniker verließ.

Die allgemeine Frustration war enorm; aber im März 1979, nach mehr als zwei Jahren, konnte man dennoch beginnen. Richardson, aus einer gewissen Vorsicht, schlug Segal vor, für die intensive Überwachung und Beratung, die sicher notwendig sein würde, einen Mitarbeiter zu nehmen. Jon Broome, der damals das Bauamt schon verlassen hatte, schien die einzige Person zu sein, die dafür in Frage kam. und Segal war glücklich, mit diesem Architekten zu arbeiten, dessen Vater vor vielen Jahren schon einmal kurze Zeit sein Assistent gewesen war. Nur einer der Selbstbauer sprang ab, und Jon Broome war froh, daß er dessen Platz übernehmen und sein eigenes Haus bauen konnte, in einer Hausgruppe, die man später »Segal Close« nannte. Mitte 1980 war alles fertig.

Brian Richardson feierte das Ereignis in einem Brief

**First Lewisham self-built project under construction.
Die erste Gruppe der Selbstbauhäuser in Lewisham im
Bau.**

replaced by an indecisive officer who relied on a
small-minded quantity surveyor.

Frustration all round was enormous; but in March
1979, work at the site finally began. Treading with care,
Richardson suggested to Segal that he needed someone to work with him in the intensive supervision and
advice which would be required; Jon Broome, who had
already left Lewisham, was the obvious, indeed the only,
person who could do this, and Segal was happy to
begin working with this architect whose father had briefly been his assistant many years before. Only one self-
builder dropped out and Jon Broome was pleased to
be able to fill the gap and build his own home, in the
group which was to be named »Segal Close.« By mid-
1980 they were complete.

Brian Richardson celebrated it in a letter to Segal:
»Well we did it. Or rather the self builders did it and
insisted on our conniving at their success. Fourteen
of the most satisfactory council houses ever built and
what is better, fourteen families imbued with a new
self-confidence and sense of control over their own
lives. Your method of building is certainly vindi-
cated.«[11]

By March 1984, though Lewisham politicans (and
also the borough's subtle political complexion) had
changed, and though the officers (now without
Richardson) had much less enthusiasm, a second
scheme was launched successfully if slowly. The
frames of these thirteen two-storey houses on one very
steeply sloping site were half-up when Segal died in October 1985; the group of homes inhabited by the end of
1986 has been named »Walter's Way.« (He is the only
living British architect ever to have London roads
named for him; more importantly, names chosen by the
residents.)

Pygmalion?

Single-handedly, Segal invented the impossible idea of
ordinary, non-skilled working-class women and men
from the waiting list for public authority housing, building their own homes. Lewisham self-builders have
ranged from retired men in their 60s to single mothers;
many are families with young children who construc-
tively joined in creating their own home.

Segal felt increasingly stimulated by his work with
self-builders, which utterly changed his own area of
control. When a self-builder told him »Walter you
couldn't do a thing without me – you may be able to

Self-builders at Lewisham.
Selbst-Bauer in Lewisham.

an Segal: »Nun, wir haben es geschafft. Oder, eigentlich, die Bauherrn selber haben es geschafft und darauf bestanden, uns an ihrem Erfolg teilhaben zu lassen. Es sind 14 der gelungensten Sozialhäuser, die je gebaut wurden, und, was noch besser ist, vierzehn Familien mit einem neuen Selbstvertrauen und einem Gefühl, ihr eigenes Leben unter Kontrolle zu haben. Ihre Methode ist sicherlich gerechtfertigt.« (Aus *Letters from a Papermakers' Husband*)

Im März 1984 konnte ein zweiter Abschnitt erfolgreich, wenn auch langsam in Angriff genommen werden, obwohl in der Zwischenzeit die Politiker in Lewisham und auch das politische Klima gewechselt hatten und obwohl die Beamten (nun ohne Richardson) viel weniger Begeisterung mitbrachten. Die Rahmen für die 13 Häuser auf einem steil abfallenden Grundstück waren zur Hälfte aufgerichtet, als Segal im Oktober 1985 starb. Die Hausgruppe, die Ende 1986 bezogen wurde, heißt »Walters' Way«. Damit ist Segal der einzige britische Architekt, nach dem zu Lebzeiten Londoner Straßen benannt wurden. Und: Es sind die einzigen Straßennamen, die die Bewohner selber gewählt haben.

Die Skala der Selbstbauer in Lewisham reichte von Rentnern über 60 bis zu alleinstehenden Müttern. Viele waren Familien mit kleinen Kindern. Alle fanden sich zusammen, um ihr eigenes Heim zu bauen. Segal fühlte sich mehr und mehr stimuliert durch die Arbeit mit den Selbstbauern, die sein Kontrollbedürfnis vollkommen veränderten. Als ein Selbstbauer ihm einmal sagte: »Walter, Du könntest nichts ohne mich machen – Du kannst vielleicht etwas zeichnen; aber Du brauchst mich, um es auszuführen«, erzählte er diese Geschichte mit offensichtlichem Vergnügen weiter und fügte hinzu: »Das ist goldrichtig, nicht wahr?« (Aus Charlotte Ellis: *Walters' Way*, The Architectural Review, März 1987, S. 77–81)

Vielleicht ist es in gewissem Sinne auch wahr, daß er die Tendenz hatte, von der Kontrolle über materielle Dinge, die er zu Bauten zusammengesetzt hatte, weiterzugehen zur Kontrolle über Menschen, die selbst bauen. Aber das ging nicht. Er konnte Menschen nicht wie Material behandeln. Er freute sich vielmehr an ihren Improvisationen, ihrer Anpassungsfähigkeit, die durch seine präzisen Regeln möglich geworden war. »Wir haben uns selbst von den Fassaden der Architekten befreit, endlich«, sagte er, wie der Erfinder eines Spiels sich an denen freut, die es beherrschen und spielen.

draw things but you need me to carry them out,« he repeated the story with evident enjoyment, adding »It's dead true, isn't it?«[12]

Perhaps in a sense it is true that he had moved from control in the material world, assembling it into building, to control of self-builders, the hands which came to life within his architecture. So he didn't fight them as he had fought tooth and nail with builders, as he had stood up to clients in the past. Segal always loved his material, »seeing through« it to reveal unexplored potential. Now the self-builders were his material. So he delighted in their improvisation and adaptation, made possible by his precise rules. »We have freed ourselves from the architect-designed facade at last« he would say, as the inventor of a game marvels at those who master and exploit it.

Segal would talk of how in the early 1960s he had tried to persuade Fred Wade, a traditional carpenter to use an electric hand drill. »Mr Segal, I'd rather stick to the methods I know!« he replied; so Segal bought him a Black & Decker drill. »I put it into his hands, switched it on for him and his eyes popped out. He was utterly astonished – and, of course, became highly skilled.« (Indeed it was this carpenter who built »the little house« and many subsequent ones.)

Similarly Segal loved to describe how Lewisham self-builders developed the system; how one, a docker, devised a way of building the walls of his two-storey house without scaffolding – a problem Segal had unsuccessfully addressed.

The family of 13 separate, similar but not identical chalets cascading down among the trees, the second Lewisham scheme, show Segal's nearest approach to that goal of equipotential space: balanced, poised in a dynamic equilibrium, speaking both to community and individuality – so reminiscent of his lovely image from 1934 of Deya in Mallorca. »Casually fitting into an existing non man-made environment, building into trees, on sloping land...« as he had said years before. This place, with the social cohesion of inhabitants involved in their place making not in shaping a made place, makes for a stronger public realm than Krier-like patterns of streets. But it is visually much less assertive, speaking to those who deserve to hear.

So in Lewisham the threads of Segal's guiding principles come together in rational, unassuming, likeable houses which semi-skilled people could fabricate and dwell in; in the processes of which they could themselves grow: self building, building selves.

Of course even »Segal method« building is hard work, especially when fitted into weekends and evenings. »We worked every evening for three months; we lost a lot of sleep.... We gained in the satisfaction and confidence in our own abilities, a certain basic satisfaction in building a ›shelter for oneself‹. I'm sure we will remember the experience forever.«[13]

It takes courage, but there is a deep certainty in the method. »His concept,« said one self-builder, »is that if he makes you sit down and think about the drawings, you will understand what to do.«[14] »He taught us to think for ourselves and gave us such confidence when we finished our houses we felt we could go on to do anything we set our minds to – he literally changed our lives.«[15]

The first to start at Lewisham, Ken Atkins, a floor-layer by trade, became a dynamic and central force in the success of the project. He began to travel with Segal and Broome to talk about the project, he helped others with their structures and was employed as consultant by builders using the »Segal method«. During the construction of frames for the Segal exhibition in London in May 1988, he said to me: »I still miss him very much. So often. He taught me so much. I want to ring him up and just chat, as he would, about anything, the world! Basically he taught you that you can conquer the world if you put your mind to it. That's what I learned from him. Changed my world.«

Eupalinos or the Architect, Paul Valéry's great poem dedicated to the power of *homo faber,* and the essence of humanity being in doing rather than theorising, takes the form of a Socratic dialogue with Phaedrus. Phaedrus uses the story of his friend Eupalinos, an architect, to convince Socrates of the importance of material form. Eupalinos suggests: »By dint of constructing .... I truly believe that I have constructed myself«. To which Valéry's Socrates replies »To construct oneself, to know oneself – are these two distinct acts or not?«[16]

To Valéry, as to Violett-le-Duc in his fable about building a house, the architect epitomises the creative man.

What Walter Segal is allowing, on the other hand,

Fig. p. 175 / Abb. S. 175 >
The second Lewisham self-built project.
Zweite Etappe der Selbstbauhäuser in Lewisham.

is for this »natural«, archetypal power of building our-selves by dint of building our edifices, to be returned from the professional domain of architects to us all, women, men and children. It is as close as practical reality has come to the motto of Lethaby, whom Segal so admired: »An artist is not a special kind of person, but every person is a special kind of artist.«

The image of the colonisation of land by ordinary people amidst the decayed, late-20th century welfare state captalism; and then seeing their self-evident skill and confidence maturing through the making of their homes, is remarkable. It allows the imagination of a new conviviality in dwelling.

For Walter Segal it was also a kind of personal homecoming. He entered a new world: a community of ordinary, working people, worldly-wise and un-presuming, whose directness, immediacy and inde-pendence must have reminded him of the Ticinese of his youth.[17] In the pivot between these, in his writings of the 1940s, Segal had clearly declared his faith in the common man's common-sense; the ground of democracy being the reasoning of the unprejudiced worker. Lewisham allowed Segal a glimpse of this world in action.

Shall we forever resign the pleasure of construction to the carpenter? What does architecture amount to in the experience of the mass of men?...
H. D. Thoreau »Walden« (1845–47)

Notes and Quotations

1. »The Uthwatt Committee recognised that land nationalisation is the most satisfactory way of dealing with the problem, but rejected it because of the political controversy it could cause.« (p150) »...we have shown how necessary it is that all land of the country should be in national ownership. This demand has support far beyond the limits of the labour move-ment.« (p173) from *Homes for the People*, co-authored by Segal, edited Andrew Boyd and Colin Penn, for the Associa-tion of Building Technicians, pub. Paul Elek, 1945.
2. His own house in St Anne's Close was typical, built for £2500 and worth, with thirty years off the lease, £250,000.
3. *The Guardian*, 2 June 1988
4. In the early 1970s, John F C Turner, who had made a name working with self-build *barridas* in Latin America, and then more widely with self-build (as seen in *Freedom to Build*, the influential book he co-wrote), settled in London. Under him, a Dweller Control Housing Group met at the Architectural As-sociation.
5. »I told him that in my view, rightly or wrongly, Camden were a bunch of Labour Party ideologists with a mission to house everyone and would see him as some kind of crank. I said I thought the only London borough where the idea stood a chance was Lewisham,« (Colin Ward, correspondence with the author, 17 February 1988), and he explained his reasons, based on the personalities of leading councillors and officers.
6. Nicholas Taylor, »Learning the Lewisham Way«, *The Architects' Journal*, 18 May 1988, p87.
7. I am grateful to Peter Stead, who was the project's midwife, and who drew my attention to and recorded it in his *Self-Built Housing groups and Co-operatives*, 1979.
8. See my »The Anarchy of Planning« (not my title), *Building Design*, 17 March 1978, pp14–15.
9. Ward, as 5. Richardson published this letter as memory of that meeting: »Dear Colin (Summer 1976) Thank you for a woncer-ful party. How you can be such a self-effacing fellow yet so calmly change people's lives for them I don't know. Nothing will be the same now you have introduced me to Walter Segal! There was I, working in a local government architects' department wondering how to build houses for people that were not distressingly like off-the-hook suits of uniform, anc there was Walter, trying to find a London borough ready for his experiment of allowing council housing applicants to build their own houses. Then, through you, we meet and a col-laboration is born that simply must be fruitful. He is a most un-usual and extraordinary man who seems to refuse to let his sensible but unorthodox ideas be baulked. What an upsetter of the Establishment! Here goes, I feel quite vertiginous. Yours fraternally Brian« (Brian Richardson, from »Letters from a Papermaker's Husband«, *Matrix*, Whittington Press; reproduced by permission of the author)
10. Brian Richardson, in conversation with the author, 28 February 1988.
11. Richardson, as in 9.
12. from Charlotte Ellis »Walter's Way«, *The Architectural Review*, March 1987, pp77–81.
13. Michael Holland, the first Segal self-builder, in letter to the author, 31 January 1975.
14. quoted in Ellis, as 12.
15. quoted in Pete Sutton, building study of 17b Longton Avenue, Polytechnic of North London, March 1987.
16. A wider context for this quotation is given in the frontispiece.
17. Colin Ward suggests a comparison between Peter Kropotkin's conversion to anarchism while among the independent-minded working people of the Jura and Segal in Lewisham.

Quotations from Walter Segal in this section from: 1945(B)1; 1971(B)2; unpublished letter to Brian Richardson, 5 October 1974, quoted with permission; 1984(C)1.

Segal sprach gerne davon, wie er einmal in den frühen Sechziger Jahren versucht hatte, Fred Wade, einen traditionellen Zimmermann, dazu zu bringen, einen elektrischen Bohrer zu benutzen. »Herr Segal«, sagte der, »ich bleibe lieber bei den Methoden, die ich kenne.« Da kaufte ihm Segal einen Black&Decker-Bohrer. »Ich gab ihn ihm in die Hände, schaltete ihn ein, und seine Augen fielen fast heraus. So erstaunt war er, und später wurde er natürlich sehr geschickt damit.« Wade war es, der das *kleine Haus* baute und später noch viele weitere.

Ebenso liebte es Segal, zu beschreiben, wie die Bauherrn in Lewisham das System entwickelten, wie Bill Gosby einen Weg fand, die Wände seines zweistöckigen Hauses ohne Gerüst zu bauen, und so weiter.

Die Gruppe von 13 freistehenden, ähnlichen, doch nicht identischen Häusern, ist Segals gelungenster Versuch, seinem Ziel eines ausgewogenen Raumes nahezukommen: Ausbalanciert, schwebend, in einem dynamischen Gleichgewicht, sowohl Gemeinschaft wie Individualität ausdrückend, erinnert die Siedlung an das beglückende Bild von Deya auf Mallorca aus dem Jahre 1934. »Zufällig eingefügt in eine nicht vom Menschen geprägte Umgebung, Bauen zwischen Bäumen, auf hügeligem Land ...« Der soziale Zusammenhalt der mit diesem Ort durch ihre eigene Arbeit verbundenen Bewohner führt zu einer viel stärkeren öffentlichen Besitzergreifung als die künstlich geformten Straßenmuster zum Beispiel eines Krier. Aber visuell ist das Ganze weniger bestimmend. Es spricht nur zu denen, die es verdienen.

So vereinen sich in Lewisham die Fäden von Segals Vorstellungen. Sie kommen zusammen in vernünftigen, bescheidenen, liebenswerten Häusern, die kaum ausgebildete Leute herstellen und bewohnen können, in einem Prozeß, an dem sie selbst wachsen konnten. Selbst-bauen, Sich-selbst-bauen.

Natürlich bedeutet auch die Segal-Methode harte Arbeit, ganz besonders, wenn sie an Feierabenden und Wochenenden erbracht werden muß. »Wir haben jeden Abend gearbeitet, drei Monate lang; wir kamen oft um unseren Schlaf ... aber wir gewannen Befriedigung und Vertrauen in unsere eigenen Fähigkeiten. Es gibt ein gewisses grundlegendes Selbstgefühl, die eigene Behausung zu bauen. Ich bin sicher, daß wir uns an diese Erfahrung immer erinnern werden.« (Michael Holland, der erste von Segals Selbstbauern, in einem Brief an den Autor, Dezember 1974)

Es braucht Mut; aber es liegt auch Sicherheit in dieser Methode. »Sein Konzept ist«, sagte ein Selbstbauer, »daß er Dich dazu bringt, Dich hinzusetzen und über die Zeichnungen nachzudenken, dann versteht man, was man tut. Er lehrte uns, selbst zu denken, und gab uns so viel Selbstvertrauen, daß wir, nachdem unsere Häuser fertig waren, das Gefühl hatten, daß wir alles tun könnten, wonach uns der Kopf stände – er änderte buchstäblich unser Leben.«

Der erste, der in Lewisham anfing, Ken Atkins, Bodenleger von Beruf, wurde eine dynamische, zentrale Figur für den Erfolg des Projektes. Er begann mit Segal und Broome zu reisen und über das Projekt zu sprechen; er half andern beim Bau und wurde als Berater bei Unternehmen angestellt, die nach der Segal-Methode bauten. Während des Zusammensetzens der Rahmen bei der Segal-Ausstellung in London im Mai 1988, sagte er zu mir: »Ich vermisse ihn sehr. So oft. Er lehrte mich so viel. Ich würde ihn so gerne anrufen, einfach mit ihm reden, so wie er es tun würde, über irgend etwas, über die Welt. Im Grunde lehrte er mich, daß man die Welt erobern kann, wenn man es sich in den Kopf setzt. Das lernte ich von ihm. Das änderte meine Welt.«

Eupalinos oder der Architekt, Paul Valérys großes Gedicht, der Essenz des Menschseins gewidmet, die darin liegt, etwas zu tun, nicht nur zu philosophieren, ist in die Form eines Dialogs zwischen Sokrates und Phaedrus gebracht. Phaedrus bedient sich der Geschichte von Eupalinos, einem Architekten, um Sokrates von der Wichtigkeit der materiellen Form zu überzeugen. »Indem ich baute, warf er lächelnd hin, habe ich mich, glaube ich, selbst erbaut.« Worauf Valérys Sokrates antwortet: »Sich erbauen und sich selbst erkennen, sind das zwei getrennte Akte oder nicht?«

Walter Segals Anliegen ist, daß die natürliche archetypische Kraft, uns selbst zu bauen, indem wir unsere Gebäude errichten, an uns alle, Frauen, Männer und Kinder, zurückgegeben wird. »Ein Künstler ist nicht ein besonderer Mensch, aber jeder Mensch ist in besonderer Art ein Künstler.«

Die Vorstellung einer Besiedlung des Landes durch gewöhnliche Menschen inmitten des zerfallenden Wohlfahrtsstaats-Kapitalismus des späten 20. Jahrhunderts ist keine Utopie. Die selbstverständliche Handfertigkeit und das Vertrauen, das reift, wenn Leute ihr Haus bauen, liefern den Beweis.

Lewisham 2

Lewisham 2

The Colour Supplement

»We are all brought up to tell the truth. Walter saw no need to. You have to *deserve* the truth, he would say; there's absolutely no question...

But we are all liars, are we not?«[1]

Julius Posener published an amusing and unexpected reminiscence: »Walter once told me that he suffered from a superstitious belief in astrology which he found somewhat hampering. He therefore undertook to climb a difficult mountain on a day which from an astrological point of view was inauspicious, even dangerous, hoping thus to liberate himself from such influences. I shudder to think of him there all by himself.«[2]

Segal on his father's knee, 1908.
Segal auf den Knien seines Vaters, 1908.

1916

Bildbeilage

Von Julius Posener eine amüsante und überraschende Erinnerung: »Walter erzählte mir einmal, daß er an einer abergläubischen Bindung an die Astrologie leide und das als lästig empfinde. Er machte sich deshalb an eine schwierige Bergbesteigung, an einem Tag, der vom Standpunkt der Astrologie aus ungünstig, ja sogar gefährlich war, in der Hoffnung, daß er sich dadurch von solchen Einflüssen befreien könne. Es schaudert mich, wenn ich an ihn dachte, da oben ganz allein.« (The Architects' Journal 4. 5. 88, S. 40)

»Unsinn«, sagte Moran Segal zu mir. »Es ist absurd zu denken, daß Walter an Astrologie glaubte!«

Davon ausgehend, daß beides wahr sein könne, gab ich Tag und Zeitpunkt der Geburt eines Unbekann-

1933

Outside ski house, 1960.
Vor dem Skihaus, 1960.

»Nonsense!« retorted Moran Segal to me. »It is absurd to think of Walter believing in astrology!«

Assuming both to be true, I gave simply the date and place of birth of an »anonymous« man to one who could prepare a chart. This is the unedited text of what was »seen«:

First of all, practical.
Competitive, aggressive; very strong and very much a sports person. Sport is heavily exaggerated.
Not academic, takes time to assimilate ideas, but once they are taken on board, they remain. Knowledge, which is wide-ranging, is assimilated through practice not theorising.
This creativity is not abstractly intellectual but very *grounded*; (Mercury in Taurus). Innovative, but not impressed by »academia« in the traditional sense.
More of a spectator.
Dynamic and very, very practical.
Nervous; lots of volatile energy.
Tremendous laziness and tremendous energies, in extreme swings. Diligent, meticulous, neat.
He lies about things because he's actually stolen it from somebody else.
Someone who strategically plans goals and heads for them, despite whatever hurt it may cause. Someone who »finishes the project«.
A man of his word. One who goes by the rules – as in sport, the clear rules of the game.
Ambitious but a bit abraisive; selfish tendencies. Certainly argumentative.
Sexual prowess emphasised; proud, but insecurities are there.
An ear for music? Someone whose activity is in a sense public (as sport or music), is for everybody – but then whose ego steps in and manages to get people's backs up. Problems with friends, and with family. Sudden endings, with both friends and family – and he's unaware why the break has happened.
Family heartbreak; considerable pain. And yet he seeks freedom.
So many elements in contradiction! Quite an extraordinary character!
What may seem to others »fun-loving« attitudes, can be explosive here.
A man's man; likes male company. Not a family man.
Has to come to terms with what could be, rather than what is.
Psychic? Probably a healer, and he surely doesn't know it.
Not an abundance of water, it's in the outer planets. Dry.«

Well? Who's going to listen to astrology. After all, Segal came down the mountain safely.

Walter Segal was very short and immensely strong. Always active in sport, he had remarkably powerful shoulders and upper arms. There are memories of his throwing much larger men to the floor. Fellow students remember him at the gym or swimming; in London colleagues remember him in early morning tennis games or the swimming pool; and always he skied. Nothing was more exhilerating than skiing alone, high up in the mountains, on a fine day, naked.

One day, skiing far up the mountains and alone, Walter was caught equi-distant from the nearest settlement and his extremely isolated ski house as night came on. »What did you do?« Segal was asked by Richard Rogers, who recounts the story. »I took off my skis, planted them in the snow fifteen metres apart, and spent the night walking from one to the other dreaming of what I would build if this were my site.«

»What made Walter« (according to a non-architect friend[3]) »so amazing a person were
– First, the breadth of his reading coupled with a perfect memory of everything he had ever read;
– Second, his *complete*, independence of thought;
– Third, his physical strength and stamina«

»The house is so compact and neat!« (marvelled a client from the 1950s)[4]. »If ever something needed replacing, like when a plumber installed a new boiler, the tradesman always turns to me and says: ›Who built this house, a dwarf?‹ And I reply: Yes!«

Of the four elemental beings (as recorded by Paracelsus but from an ancient mist), the personification of earth was the dwarf. The dwarf is grounded, his cleverness is his feet-on-the-ground reality. He is cunning and hard-working, potent and immensely strong – as the arms and shoulders of the ancient Egyptian dwarf Bes.

I remember a meal on a sunny day in my little London house with him and a mutual Swiss friend, eating bread, coppa and parmigiano cheese and drinking red wine. Three men laughing at the foibles of our favourite jokers, the architectural butterfly collectors and pastrycooks.

I remember a meal Marion and I had with Walter and Moran and another couple one warm summer evening on their verandah, next to the quietly ageing »tem-

ten an jemanden, der ein Horoskop aufstellen kann. Ich zitiere, unbearbeitet, was dabei herauskam:

Zunächst das Wichtigste: praktisch.
Konkurrenzfähig, aggressiv; sehr stark und sehr sportlich, übertrieben sportlich.
Nicht akademisch, braucht Zeit, um Ideen aufzunehmen. Sind sie jedoch einmal aufgenommen, bleiben sie. Umfassende Kenntnisse, mehr durch die Praxis als theoretisch erworben.
Eine Kreativität, die nicht intellektuell abstrakt ist, sondern auf dem Boden der Tatsachen steht (Merkur im Stier). Innovativ, aber nicht beeindruckt durch »accademia« im traditionellen Sinn.
Eher ein Zuschauer.
Dynamisch und sehr, sehr praktisch.
Nervös, eine Menge sprunghafte Energie.
Ungeheure Trägheit und ungeheure Energien im extremen Wechsel.
Fleißig, genau, sauber.
Er lügt über Dinge, weil er sie tatsächlich von jemand anderem gestohlen hat.
Jemand, der seine Ziele strategisch plant und auf sie zugeht, egal was es kostet. Jemand, der sein Projekt zu Ende bringt.
Ein Mann, ein Wort. Einer, der den Regeln folgt, wie im Sport, den klaren Regeln des Spiels.
Ehrgeizig, aber ein bißchen zermürbend; selbstsüchtige Tendenzen. Sicher auch streitsüchtig.
Sexuell angriffig, stolz, aber Unsicherheiten sind auch da.
Ein Ohr für die Musik? Jemand, der in einem gewissen Sinne für die Öffentlichkeit da ist, für jedermann (Sport, Musik) – aber dessen Ego dazwischentritt und Menschen gegen sich aufbringt.
Probleme mit Freunden und mit der Familie. Ein plötzliches Ende, sowohl mit Freunden als auch mit der Familie – und er weiß dann nicht, warum.
Gebrochene Herzen in der Familie, großer Schmerz, und doch sucht er die Freiheit.
So viele gegensätzliche Elemente. Ein ganz außergewöhnlicher Charakter.
Was andern als Freude am Spaß vorkommt, kann hier explosiv werden.
Ein Mann für Männer; liebt männliche Gesellschaft.
Muß in Einklang kommen mit dem, was sein könnte, nicht dem, was ist.
Psychisch? Wahrscheinlich ein Heiler; aber sicher weiß er das selber nicht.
Keine Rührung, das ist außerhalb seiner Sphäre; trocken.

Nun? Wer hört auf Astrologie? Auf jeden Fall kehrte er heil von seiner Bergtour zurück.

Walter Segal war klein und ungeheuer stark. Er war immer sportlich aktiv, hatte bemerkenswert kräftige Schultern und Oberarme. Mitstudenten erinnern sich an ihn in der Turnhalle und im Schwimmbad; Londoner Kollegen erinnern sich an Tennisspiel und Schwimmen am frühen Morgen. Und immer ist er Ski gefahren. An nichts hatte er mehr Freude als am Skifahren, allein, hoch oben in den Bergen, an einem schönen Tag, nackt.

Eines Tages hoch in den Bergen, wurde er von der Nacht überrascht, gleich weit entfernt von der nächsten Siedlung wie von seinem einsamen Skihaus. Richard Rogers, der die Geschichte erzählt, fragte: »Was hast Du gemacht?« »Ich nahm meine Skier ab, pflanzte sie etwa 15m voneinander entfernt in den Schnee und bin die ganze Nacht von einem zum andern hin und her gelaufen und habe davon geträumt, was ich bauen würde, wenn das mein Grundstück wäre.«

»Was machte Walter zu der erstaunlichen Persönlichkeit, die er war? – Zunächst einmal die Breite seiner Lektüren, zusammen mit dem perfekten Gedächtnis für alles, was er je gelesen hatte. Zweitens, die absolute Unabhängigkeit seines Denkens. Drittens, seine physische Kraft und Ausdauer.« (Sein Freund John Allsopp, nicht Architekt, in einem Brief an den Autor, März 1988)

Ein Bauherr aus den Fünfziger Jahren berichtet: »Das Haus ist so kompakt und akkurat. Wenn etwas ersetzt werden muß, zum Beispiel der Klempner einen neuen Boiler bringt, dreht er sich um zu mir und fragt: Wer hat dieses Haus gebaut, ein Zwerg? Und ich antworte: Ja.« (Monica Pidgeon im Gespräch mit dem Autor, März 88)

Von den vier elementaren Wesen (nach Paracelsus, aber beruhend auf einem alten Mythos) ist der Zwerg die Personifikation der Erde. Der Zwerg steht fest auf dem Boden, ist schlau und realistisch, ist geschickt, arbeitet hart, ist potent und ungeheuer stark – wie der alte ägyptische Zwerg Bes.

Wir wurden alle dazu erzogen, die Wahrheit zu sagen. Segal hielt das nicht für nötig. Man muß sich die Wahrheit verdienen, sagte er. Es ist gar keine Frage ... Aber wir sind doch alle Lügner, nicht wahr? (Moran Segal in einem Gespräch mit dem Autor, April 1988)

Ich erinnere mich an eine Mahlzeit an einem sonnigen Tag, in meinem kleinen Londoner Haus, mit einem gemeinsamen Schweizer Freund. Wir aßen Brot, Coppa und Parmesan und tranken roten Wein. Drei Männer, die über die Schwächen ihrer alten Lieblingswitze lachten, drei Schmetterlings-Sammler und Zuckerbäcker.

porary« house and looking up at the looming shadow of Berthold Lubetkin's Highpoint.

I remember a grey, watery snowstorm in a venerable, beautifully chunky and enclosing timber chalet, down the hill from the fashionable Swiss ski resort of Verbiers; all cooped-up and humid; playing games, talking and drinking too much excellent Poire William spirit. (For the night before, we had also drunk much, round a tiny table in a tiny Corbu duplex apartment in maison Clarté in Genève, with Walter's scurrilous and hilarious tales behind the scenes in the 1930s.) Now Walter, all the while, was fretting that it was too wet and soft to ski. He was in his seventies.

Throughout the 1960s, *The Architects' Journal* used Walter as »Astragal's learned spy« and as book reviewer, to write a vast number of notes, reviews and on exhibition visits, all the while displaying his immensely wide reading and amazing visual memory. He corrected historians, arguing with Rudofsky, ticking off Banham, and gleefully enjoying himself. His review of the *Penguin Dictionary of Architecture*, for example, is an amazing display of vast-ranging erudition, wrapped

The shadow of the architect.
Der Schatten des Architekten.

The feet of the architect and Swiss snow.
Die Füße des Architekten mit Schweizer Schnee.

round in modesty. In a *Dictionary of Penguin Architecture*, of course, Segal would have been one of the few entries.

The Times obituary entitled Walter Segal: »Teacher of Architecture.«[5] It began »He achieved a unique position within the profession as an individualist with convictions about the proper relationship between architect, client and builder which he put forward indefatigably and persuasively .... It made him an out-of-the-ordinary teacher.«

Yes.

»A Swiss architect,« it continued, »Walter Segal was born at Ascona,... and was granted his diploma at the Technische Hochschule Zürich in 1932.«

Equally true, but utterly imaginal.

He wrote suggesting Switzerland as his home[6], and other carefully chosen references reinforced the implication[7]. Close friends from that time (Monica Pidgeon and Colin Boyne) never doubt it even today; »But the Swiss are born on skis, aren't they?« says one. »Ah, he's the most influential Swiss in the history of British architecture ! It's true isn't it?« adds another. (These two are perhaps the most influential architectural editors in post-war Britain.[8])

At least, to recall another Helvetian, he didn't hide his seagull under a crow.

»I will never tell lies, though I cannot promise always to tell the whole truth.«

So said Hermes[9] just after he had told lie after lie to Apollo and Zeus about the theft of cattle.

»The accidental falling-into-your-lap as the Hermetic material; its transformation through finding-thieving (the Hermetic event) into a Hermetic work of art which is always something of an optical illusion....

»Hermes, sometimes old and bearded, sometimes youthful; this youthfulness remains always characteristic of him, the prototype of a playfully and nibmly unfolding masculinity....

»Despite the shamelessness – and this is probably the most wonderful thing about it – a divine innocence is properly suited to and inherent in the Hermes world. Hermes has nothing to do with sin and atonement. What he brings with him from the springs of creation is precisely the innocence of becoming.«[10]

These are the final words of Karl Kerenyi's *Hermes Guide of Souls*. Kerenyi closes the Monte Verità circle; his grave in Ascona has the inscription used in antiquity for those »initiated into the mysteries of Hermes«: *tetelesmenoi Hermei*

Walter's curiosity was insatiable; the length of his telephone calls, on any subject under the sun, legendary.

»Walter,«recalls an institutional client for self-build houses, »had a distractingly puckish exterior – the domed head bobbing up and down, the false teeth falling over themselves with enthusiasm – yet what I remember most lastingly about him was his perfect courtesy to all, his high expectations of everyone and the way in which he exemplified those virtues of real equality that his self-build houses expressed on the ground.«[11]

In her memorial note[12], Charlotte Ellis quoted a discipline which Walter Segal encouraged on many friends: »You must resist the impulse to tell so much truth, learn to procrastinate then, when least expected, act with lighting speed.«

»Shameless trickster, divine scoundrel all wrapped in cunning wiles....« (Homeric Hymn to Hermes)

Notes

1. Moran Segal, in conversation with the author, April 1988.
2. *The Architects' Journal*, 4 May 1988, p40.
3. John Allsopp, letter to the author, March 1988.
4. Monica Pidgeon, in conversation with the author, March 1988.
5. *The Times*, 2 November 1985.
6. as in *The Architect & Building News*, August 1949.
7. *The Architect & Building News*, April 1955 is a good example.
8. Monica Pidgeon, at *Architectural Design* and latterly *The Journal of the Royal Institute of British Architects*; Colin Boyne at *The Architects' Journal* and latterly also *The Architectural Review*.
9. the words of Robert Graves, *The Greek Myths*, 1985, p26.
10. Karl Kerenyi, *Hermes, Guide of Souls*, 1976.
11. Nicholas Taylor, »Learning the Lewisham way«, *The Architects' Journal*, 18 May 1988.
12. *The Architects' Journal*, 6 November 1985, p28.

Ich erinnere mich an eine Mahlzeit, die Marion und ich mit Walter und Moran an einem warmen Sommerabend auf ihrer Terrasse, gerade neben dem ruhig alternden temporären Haus einnahmen, mit dem Blick auf die drohenden Schatten von Berthold Lubetkins Highpoint.

Ich erinnere mich an einen grauen, wäßrigen Schneesturm. Wir waren in einem ehrwürdigen kompakten und gemütlichen Chalet unterhalb des Schweizer Nobelkurortes Verbier, alle eng beisammen, spielten, redeten und tranken zu viel von dem ausgezeichneten Williams-Schnaps. Walter hat sich die ganze Zeit geärgert, daß es zu naß und zu weich zum Skilaufen sei. Er war schon über siebzig. (In der Nacht zuvor hatten wir auch viel getrunken, rund um einen winzigen Tisch in einem winzigen Corbusier-Duplexappartement in der *Maison Clarté* in Genf, bei Walters skurrilen und ausgelassenen Geschichten.)

Durch die Sechziger Jahre hindurch schrieb Walter für das Architects' Journal als »Astragals gelehrter Spitzel« eine große Zahl von Notizen, Buch- und Ausstellungsbesprechungen, und legte dabei immer eine ungeheuer weit gefächerte Belesenheit und sein verblüffendes visuelles Gedächtnis an den Tag. Er korrigierte Historiker, stritt sich mit Rudofsky, wischte Banham eins aus und genoß das alles sehr. Seine Besprechung des *Penguin Dictionary of Architecture*, zum Beispiel, zeigt ein ganz erstaunlich umfassendes Wissen, in Bescheidenheit verpackt. In einem *Dictionary of Penguin Architecture* wäre Segal wohl einer der wenigen gewesen, die man hätte nennen können.

Der Nachruf in der »Times« bezeichnete Segal als Architektur-Lehrer: »Er erreichte eine einmalige Position unter den Architekten als ein Individualist mit festen Ansichten über die Beziehung zwischen Architekt, Bauherr und Baumeister. Seine Meinungen trug er unermüdlich und mit Überzeugung vor. (...) Das machte ihn zu einem ungewöhnlichen Lehrer.«

Ja.

»Der Schweizer Architekt Walter Segal«, ging es weiter, »wurde in Ascona geboren (...) und erhielt sein Diplom an der ETH in Zürich 1932.«

Auch wahr, aber ein absolutes Märchen, wahrscheinlich von ihm selber so lange und unermüdlich erzählt, bis es alle glaubten. Wenn er schrieb, stellte er

immer die Schweiz als seine Heimat dar, und andere sorgfältig gewählte Bemerkungen erhärteten diese Vorstellung. Enge Freunde aus jener Zeit haben es bis heute niemals bezweifelt. »Aber die Schweizer werden auf Skiern geboren, stimmt das nicht?« sagt man auch. »Er ist der einflußreichste Schweizer in der Geschichte der britischen Architektur. Das ist doch wahr?« fügt ein anderer hinzu.

Unmittelbar nachdem er Apollo und Zeus Lüge auf Lüge über den Diebstahl der Rinder aufgetischt hatte, sagte Hermes: »Ich will niemals lügen, obwohl ich nicht versprechen kann, die ganze Wahrheit zu erzählen.« (Robert Graves, The Greek Myths, 1985, S. 26)

»Der Fund als Zufall ist an sich noch nicht hermetisch, nur Stoff zu hermetischem Werk, welches aus ihm im Geist Gottes gestaltet wird.« (Karl Kerenyi: *Hermes der Seelenführer*)

»Du Gaukler, Du Tausendkünstler und Tischfreund, fünfzig Kühe ist wert, was Du da eben ersonnen...« (*Homerische Hymne an Hermes*)

»Er steht da, ob jugendlich oder bärtig dargestellt, als ein uns sonderbar anmutender kabirischer Eros: der wirkend und anschaulich gegenwärtige Ursprung und zugleich das Urbild der sich spielerisch-beweglich entfaltenden Männlichkeit.« (Karl Kerenyi: *Hermes der Seelenführer*, S. 92)

»Ihr eignet trotz Raub und Trug und Schamlosigkeit – und das ist wohl das wunderbarste an ihr – eine göttliche Unschuld. Dieser Hermes hat nichts mit Sünde oder Sühne zu tun. Was er aus der Quelle des Werdens mit sich bringt, ist eben dies: die Unschuld des Werdens.« Das sind die Schlußworte von Karl Kerenyis *Hermes der Seelenführer*. Kerenyi beschließt den Monte Verità-Kreis. Sein Grab in Ascona trägt die Inschrift, die in der Antike für die gewählt wurde, die in die Mysterien von Hermes eingeweiht sind: Tetelesmenoi Hermei.

»Walter hatte ein ausgesprochen gnomenhaftes Äußeres – der gewölbte Kopf, auf und ab sich bewegend, die falschen Zähne, fast stolpernd vor Begeisterung. Aber meine bleibendste Erinnerung an ihn war seine perfekte Höflichkeit zu allen, die hohen Erwartungen, die er in einen jeden setzte, und die Art, in der er selbst ein Beispiel für jene Tugenden der Ausgewogenheit war, die seine Selbstbauhäuser ausdrücken.« (Nicholas Taylor, *Learning the Lewisham Way*, The Architects' Journal 18. 5. 88)

Potential Development

Material for future building

Segal was fascinated by the products of the building industry – from his earliest published statement »Why we
use Heraklith« (woodwool slabs), in *Heraklith
Rundschau*, on a 1931 student project, to his praise for
Glasal (autoclaved compressed asbestos sheets), in *ac
Revue*, fifty years later. His insatiable curiosity, which
helped to »[inform himself] of the market in every way«

as he said, led him to reject many well-publicised
products and to chase the possibilties of others. »We,
who have been living in a tool revolution now for more
than 150 years or so, have acquired in the small world of
art and architecture the habit of speaking in a rather silly
way about *TECHNOLOGY* of which we are merely the ig-
norant recipient.« [The uppercase is his.] He understood
the architect's relation to the industry with unusual clarity.

**The Walter Segal exhibition at the Royal Festival Hall, London, 1988. The exhibition was designed by Florian
Beigel with students, and built by students, directed by Lewisham self-builders.
Die Walter-Segal-Ausstellung in der Royal Festival Hall, London, 1988. Die Ausstellung war konzipiert von Florian
Beigel und wurde von Studenten unter Anleitung von Selbst-Bauern aus Lewisham aufgebaut.**

Mögliche Weiterentwicklung

Materialien für zukünftige Bauten

Walter Segal war fasziniert von den Produkten der Baustoffindustrie – von seiner ersten veröffentlichten Äußerung *Warum wir Heraklith benutzen* in der *Heraklith-Rundschau*, bei einem Studenten-Projekt, 1931, bis zu seinem Lob für Glasal (hochgepreßte, autoklavierte Asbestzement-Platten), in der *ac-Revue*, 50 Jahre später. Seine unstillbare Neugier, »sich selbst auf jede Art und Weise am Markt zu orientieren«, wie er sagte, führte dazu, daß er manches weiterhin gepriesene Produkt ablehnte, und den Möglichkeiten anderer nachging.

»Wir, die wir nun seit mehr als 150 Jahren eine Revolution der Werkzeuge erleben, haben in der kleinen Welt von Kunst und Architektur die Gewohnheit angenommen, in einer recht albernen Weise über TECHNO-

Houses and community hall built for London Borough of Camden in 8 weeks, 1987. Designed by Jon Broome.
Häuser und Gemeinschaftszentrum in Camden, London, entworfen von Jon Broome, gebaut 1987 in 8 Wochen.

ANYONE CAN BUILD THEMSELVES A HOME

Are you aware of the possibilities?

The Walter Segal Trust. Cover of brochure, showing detail from second Lewisham project.
Der »Walter Segal Trust«. Deckblatt der Informations-broschüre. Es zeigt ein Detail der zweiten Bauetappe in Lewisham.

»There is no doubt that my buildings are *extreme-ly* transitional,« he said in 1984. »I am very much aware that the materials will be superseded all the time.« Convinced that for housing the future was in »assembling« rather than »building« as we know it, and seeing market pressure force components towards a 1200mm module, Segal looked forward to a more free and responsive variety of material and its upgrading by replacement.

»My way of building is not a timber system, it is a method of thinking,« he kept saying. In the short term – within a decade, he suggested in 1984 – metal will be able to replace timber frames, as safe laser hand-tools become available. In the longer term he saw support structures themselves being abandoned as safe and easily worked, non-toxic dense foams became both support and skin.

Segal's faith remained in man-made materials. He has been linked with small-scale, gentle technology; indeed he enjoyed involvement with the Centre for Alternative Technology in his last years, and another »Segal« building was erected there in 1987. But the central issue for Segal was *appropriateness*. »There is no need to use the complete potential of the western world in high technology for purposes which can be answered with medium and low technologies. We really do not need this high degree of perfection for trivial purposes like building dwellings. Valuable resources should not be frittered away.«

On the other hand, his »animal confidence« stopped his being swept by the »green« wind and its seductive finale of subsumption in Gaia. Asked if his potential materials are not expensive in fossil fuel energy for conversion to building components: »Nature is wasteful herself,« he thoughtfully replied. »It depends on how one defines waste. A complex, highly endowed and beautiful creature in the end serves only to feed another: since this can be achieved by simpler means, what was the purpose of creating the creature in the first place? We, as nature's children, are of course wasteful. But we should not forget that conservation is a human concept no matter how we try to read it into the supposed intentions of nature. And while conservation is eminently sensible, let us accompany it by exploration lest we lose our sense of balance.«

LOGIE zu sprechen, deren unwissende Verbraucher wir sind.« (Die Großbuchstaben sind von ihm.) Er sah die Beziehung des Architekten zur Industrie mit ungewöhnlicher Klarheit.

»Es besteht kein Zweifel, daß meine Bauten im Übergang zu sehen sind«, sagte er 1984. »Ich bin mir absolut der Tatsache bewußt, daß meine Materialien laufend ersetzt werden.« Überzeugt, daß für den Wohnungsbau die Zukunft im Zusammensetzen, nicht im Bauen nach alter Manier liege, und unter dem Eindruck, daß der Druck des Marktes zum 1200mm-Modul der Komponenten drängen wird, erwartete er eine freiere und verantwortungsbewußtere Vielfalt der Materialien, und durch das Ersetzen der bisherigen ständige Verbesserungen.

»Meine Art zu bauen ist nicht ein Holzsystem, es ist eine Methode zu denken«, sagte er immer wieder. Kurzfristig, innerhalb eines Jahrzehnts, meinte er 1984, wird man die Holzrahmen durch Metall ersetzen können, und es werden sichere Laser-Handwerkzeuge zu bekommen sein. Langfristig, glaubte er, würden die Tragkonstruktionen ganz verschwinden, weil dann sichere und leicht zu bearbeitende, nicht giftige dichte Schaumstoffe sowohl das Skelett als auch die Hülle bilden werden.

Segal glaubte an künstliche, vom Menschen geschaffene Materialien. Er war der kleinmaßstäblichen sanften Technologie verbunden. Seine Verbindung zum Zentrum für alternative Technologie während seiner letzten Lebensjahre machte ihm Freude, und dort wurde 1987 auch ein Segal-Gebäude errichtet. Aber sein zentrales Anliegen war immer die Angemessenheit. »Es ist nicht notwendig, das gesamte Potential der Westlichen Welt an Hochtechnologie für Zwecke aufzubieten, die man mit mittlerer oder einfacher Technologie erfüllen kann. Wir brauchen keinen hohen Grad der Perfektion für triviale Vorhaben, wie das Bauen von Wohnungen. Wertvolle Ressourcen sollten nicht verzettelt werden.«

Andrerseits hat sein animalisches Vertrauen ihn davon abgehalten, sich vom grünen Wind hinwegfegen zu lassen. Wenn man ihn fragt, ob seine künftigen Materialien nicht aufwendig im fossilen Energieverbrauch für die Herstellung der Bauteile seien, antwortet er überlegt: »Die Natur selbst ist verschwenderisch. Es kommt immer darauf an, wie man Verschwendung definiert. Eine komplexe, wohlausgestattete schöne Kreatur dient am Ende nur dazu, eine andere zu ernähren. Da

Hostel at Stuttgart University, built by students under Peter Sulzer and Peter Hübner, 1981/82.
Studentenhaus in Stuttgart, im Selbstbau erstellt von Studenten unter der Anleitung von Peter Sulzer und Peter Hübner, 1981/82.

Systems for future builders

It was surprising how imitation of his method, that surest form of flattery, never really developed in Segal's lifetime. There were a few mid-1970s projects, notably Tony Cooper's student village near East Grinstead (which sadly was not built)[1] and David Lea's lovely collection of shelterd housing at Churt[2].

»Segal« self-build was touched on by groups at English and Irish architecture schools, but more interesting possibilties were developed under Peter Hubner and Peter Sulzer into widely varied student housing in Stuttgart in the early 1980s.

The Segal system as a kit of actual parts can continue to be used with increasing refinement and little basic change. Segal's surviving colleague Jon Broome, amidst varied other work, does just that, often helped by Lewisham self-builders now employed as carpenters or consultants to a builder. A few small buildings around London in 1987-8, houses, a community hall and a small office, testify to the continuing stream.

There were at least two housing authorities interested in his method in mid-1988; the insensitive proposal of one, for 100 »Segal self-build« houses in a flat empty site intriguingly raises new social problems. The Walter Segal Self-Build Trust[3] is now set up to encourage and advise on such activity.

The future professionals

The »Segal method« could have far-reaching implications on the role of the workers in the house building industry, on attitudes to change, alteration and repair of houses, and on many other aspects. Perhaps most of all, it raises questions of housing choice and user control; it questions the deep prejudice that we are »so completely dependent on the housing supply system that we find it hard to believe that people *can* house themselves.«[4] But what of architects?

There is obvious inspiration in individual searches for authentic forms of architectural practice. Way back in the 1960s Segal sensed »the immense dissatisfaction with which young architects are seething in an age of abundance which has melted modern architecture into eclecticism. The passionate desire for a way out, for more scope and fulfillment is only too evident....« Segal's response was to exemplify a two-fold role for architects. First, this role must be one of direct technical competence: »The architect might be weaned from the superstition that visual matters are his chief concern. He might be saved from this obsession which is his instinctive preoccupation to the level of neurosis. When our buildings look natural again they will be able to appear less assertive and in a wider sense unobtrusive because they can afford to.« Second, this two-fold role must be one of social sensitivity: the housing architect must directly enable clients, who instinctively resist generosity and condescension, to help themselves in dwelling. However much the building industry is allowed to decline, Segal argued in 1982, »the need for structures for people to live in is growing all the time; and, fortunately, there are ways to meet the demand. With such a small labour force, I believe quite simply that the principle of self-help will *have* to be introduced on a much more general scale.

»It should fire the imagination, provided narrow solutions can be avoided. It will put an end to the esoteric debate raging on architecture at the moment (which is only the result of under-employment). But what is to be done with, shall we say, at least one generation of architects who are utterly ill-equipped for this new kind of role?«

So the twin concerns, Segal's earliest goals of »natural building« and »convivial place making«, resonate half a century later as themes for a future authentic practice.

Notes and Quotations

1. Cooper assisted Segal on the Godfrey House in 1971, and has built other small »Segal method« works.
2. Lea also built a vicarage and parish hall in Melling using a similar method.
3. Despite its shoe-string operation, surely many seeds will germinate; its potential was recognised in 1987 by the receipt of a UK »Community Enterprise Award« and, more remarkably, a UN »Building and Social Housing Foundation Award«, the first not to have gone to a »developing« country. The Walter Segal Self-Build Trust can be reached at P.O.Box 542, London SE1 1TX.
4. Colin Ward, *When We Build Again*, 1984.

Quotations from Walter Segal in this section from: 1973(B)3; Wills & Yeo interview 1984; 1984(C)1; 1974(B)1; 1970(B)2; *AJ* 22 June 1966, p.1510; 1969 (B).

man das auch mit einfacheren Mitteln erreichen könnte: Warum dann zuerst die Kreatur erschaffen? Wir, als Kinder der Natur, sind natürlich verschwenderisch. Aber wir sollten nicht vergessen: Auch die Sparsamkeit ist ein Konzept des Menschen, egal wie wir versuchen, sie im Rahmen des allgemeinen Planes der Natur zu verstehen. Und da der Prozeß des Sparens ungeheuer sensibel ist, laßt uns ihm durch Entdeckungen weiterhelfen, damit wir nicht unsern Sinn für das Gleichgewicht verlieren.«

Ein System für künftige Baumeister

Es ist überraschend, daß es zu Segals Lebzeiten kaum Nachahmer gab, ist das doch die zuverläßigste Form der Schmeichelei. Es gab ein paar Projekte Mitte der Siebziger Jahre in England: Bemerkenswert war Tony Coopers Studentendorf bei East Grinstead (das leider nicht gebaut wurde). Cooper assistierte Segal 1971 beim Godfrey House, später bei kleineren Arbeiten mit der Segal-Methode. Bemerkenswert ist David Leas schöne Wohngruppe in Churt. In ähnlicher Methode baute Lea auch Pfarrhaus und Gemeindesaal in Melling.

Studentengruppen an englischen, irischen und deutschen Architekturschulen befaßten sich mit Segals Selbstbau. Interessantere Möglichkeiten wurden unter Peter Hübner und Peter Sulzer für vielfältige Studentenwohnungen in Stuttgart in den frühen Achtziger Jahren erarbeitet.

Das Segal-System gewissermaßen als Grundmethode bleibt gültig, und mit Verfeinerungen weiter anwendbar. Segals Kollege Jon Broome arbeitet mit der Methode weiter, oft mit Hilfe von Lewisham-Selbstbauern, die heute als Zimmerleute oder Berater bei Baumeistern beschäftigt sind. Ein paar kleine Gebäude um London herum, entstanden 1987–88, Wohnhäuser, ein Gemeindehaus und ein kleines Bürogebäude sind Zeugen.

Jetzt, Ende der Achtziger Jahre, gibt es zumindest zwei Wohnungsbehörden, die an Segal-Selbstbau interessiert sind. Die eine wünscht 100 Segal-Selbstbauhäuser auf einem flachen Landstück, was interessante neue Probleme stellt. Der *Walter Segal Self-Build Trust* (PO Box 542, London SE1 1TX) wurde jetzt gebildet, um solche Vorhaben zu ermutigen und zu beraten.

Die Zukunft des Architekten-Berufs

Die Segal-Methode könnte weitreichende Auswirkungen auf alle haben, die im Wohnungsbau arbeiten. Änderungen und Reparaturen können von den Bewohnern selbst ausgeführt werden, größere Umbauten sogar. Vor allem aber stellt Segals Baumethode ein tief sitzendes Vorurteil in Frage: Daß wir »so vollkommen abhängig seien vom Angebotssystem auf dem Wohnungsmarkt und daß wir es kaum glauben können, daß Menschen in der Lage sein sollten, selbst für ihre Behausung zu sorgen.« (Colin Ward, *When we Build Again*, 1984)

Schon in den Sechziger Jahren spürte Segal »die Unzufriedenheit, die unter jungen Architekten gärte, daß in einer Zeit des Überflusses die moderne Architektur zu einem Eklektizismus verkommen ist. Der leidenschaftliche Wunsch nach einem Ausweg, nach größeren Chancen und Erfüllung ist nur zu offensichtlich.« Segals Antwort war sein eigenes Beispiel einer doppelten Rolle für den Architekten. Zunächst muß man technisch kompetent und direkt sein: »Der Architekt sollte von dem Aberglauben entwöhnt werden, daß visuelle Dinge sein erstes Anliegen seien. Er könnte gerettet werden von dieser Besessenheit, die zu einer instinktiven Voreingenommenheit bis zur Neurose führen kann. Wenn unsere Bauten wieder natürlich aussehen, werden sie wieder weniger bestimmend und im weiteren Sinne bescheiden wirken, weil sie sich das leisten können.«

Zweitens gehört zu dieser doppelten Rolle eine soziale Sensibilität: Der Wohnbauarchitekt muß den Klienten, die sich instinktiv gegen Herablassung sträuben, helfen, zu einer Behausung zu kommen. »Soweit man auch die Bauindustrie hat herunterkommen lassen«, beklagte sich Segal 1982, »ist doch der Bedarf an Bauten, in denen Menschen leben können, ständig gewachsen, und glücklicherweise kann man diesen Anforderungen entsprechen. Bei so wenig Arbeitskräften glaube ich einfach, daß das Prinzip des Selbst-Bauens in einem viel größeren Maßstab eingeführt werden muß. Es sollte die Vorstellungskraft anfeuern, vorausgesetzt, allzu karge Lösungen können vermieden werden. Das wird auch den esoterischen Debatten ein Ende setzen, die zur Zeit in Architektenkreisen toben (und die nur das Ergebnis der Unterbeschäftigung sind). Aber was soll man tun mit, sagen wir einmal, einer Generation von Architekten, die absolut nicht für diese neue Rolle ausgerüstet ist?«

David Lea: Houses at Churt
David Lea: Wohnhäuser in Churt.

David Lea: Vicarage and community center at Melling.
David Lea: Pfarrhaus und Gemeindesaal in Melling.

The Goal of Conviviality

The guiding principle, a flame which all his life warmed and illuminated Segal's relentless rationality, was the goal of an architecture of conviviality and gregariousness.

»His [le Corbusier's] amenity was not the amenity of upholstered and cosy comfort, nor, indeed was it austerity in the common sense. He dreamt of light, of the warmth of the sun, of an unforgettable view.... Purpose and performance can always be attended to by those who come after with great assiduity, because this is their *raison d'être;* the task of those who come first is to transmit insight and show that it is feasible and Le Corbusier was such a man.«

Perhaps Walter Segal was himself that rare being who, within his carefully chosen frame, could not only attend to purpose and performance, but also keep sight

Segal family in ski house, Fideris, Switzerland.
Die Familie Segal im Skihaus in Fideris, Schweiz.

of »the dream of light, the warmth of the sun, the unforgettable view.«

He had urged, as we saw, that studying and understanding human behaviour was the essential intellectual task of the architect. In the past, he argued, in contained and hierarchical societies, built symbolism responded to that context. But today, in our pluralist world-culture, such symbolism is inappropriate and its usage inauthentic.

But beyond just the practical and the good-looking, architecture must both welcome the body and provide a backdrop for gregariousness and contemplation, for stimulation and ease. This goal »lifts formal expression above arbitrary levels that in themselves can never be free from the fleeting sensations of visual pleasure or dislike.«

»People congregate for stimulation, to eat, to talk, to listen, to express and present themselves. They need an atmosphere, congenial, complementary, encouraging and enveloping, in which to enjoy such experiences. The accent lies on people. They must dominate, become aware of each other's presence, be stimulated to make contact.« Segal goes on to argue that »successful envelopes for gregarious activities depend on the creation of an illusion of space in which the contours are dim. Where people assemble, the architecture must recede, become a back-cloth, disappear; but it must leave an illusion. Therefore the envelope made to house gregarious activity should be under-defined, so that it may be extended by suggestion rather than in reality; that the apparent space may generate that sensation of ease in which gregarious activities prosper.« He talks of the need for intimacy in the midst of gregarious activity – a couple at a table on the edge of a dance floor, is his example – and of the essential corollary of gregarious space, the need for space into which to withdraw or retreat. Segal urged housing architects carefully to study these needs of potential clients, for without allowing space for separation (for quiet and for contemplation or for TV and for music), one can-

Heiterkeit

Die Flamme, die sein ganzes Leben lang seine unbarmherzige Vernunft wärmte und erhellte, war das Ziel einer Architektur der Heiterkeit und der Geselligkeit

»Seine Annehmlichkeiten waren nicht die der gepolsterten und gemütlichen Bequemlichkeiten, noch war es Nüchternheit im üblichen Sinne. Er träumte vom Licht, der Wärme der Sonne, einer unvergeßlichen Aussicht ... Die Erfüllung von Vorsätzen kann immer durch die erreicht werden, die mit großer Beharrlichkeit daran gehen; denn es ist ihre *raison d'être*; die Aufgabe derer, die zuerst kommen, ist es, Einsichten zu vermitteln und die Überzeugung, daß sie ausführbar sind. Le Corbusier war solch ein Mann.« (Segal über Corbusier 1965)

Vielleicht war Walter Segal selbst einer dieser seltenen Menschen, denen es gelingt, innerhalb eines sorgsam gewählten Rahmens nicht nur Vorsatz und Erfüllung in Einklang zu bringen, sondern auch »den Traum vom Licht, der Wärme der Sonne, der unvergeßlichen Aussicht« im Auge zu behalten.

Er hatte darauf gedrängt, das Studium und das Verständnis des menschlichen Verhaltens als grundlegende intellektuelle Aufgabe des Architekten zu werten. In der Vergangenheit, so argumentierte er, in geschlossenen und hierarchischen Gesellschaften, erfüllte gebaute Symbolik dieses Bedürfnis. Aber heute, in unserer pluralistischen Welt-Kultur sind solche Symbolismen unangemessen und unglaubwürdig ...

Aber über das Praktische und visuell Angenehme hinaus muß die Architektur den Körper willkommen heißen und einen Rahmen für Geselligkeit und Einkehr bieten, zur Anregung und zur Entspannung. »Menschen kommen zusammen, um Anregung zu suchen, um zu essen, zu reden, zuzuhören, um sich selbst auszudrücken und sich zu zeigen. Sie brauchen eine geistesverwandte, sie ergänzende, ermutigende und umhüllende Atmosphäre, in der sie diese Erfahrungen genießen können. Die Betonung liegt auf dem Menschen. Sie müssen dominieren, sich der Gegenwart der anderen bewußt sein, dazu angeregt werden, Kontakte zu knüpfen.«

Segal geht weiter in seinem Argument, daß »der

Erfolg einer Hülle für gesellige Aktivitäten darauf beruht, eine Illusion des Raumes zu schaffen, in der alle Konturen verschwimmen. Dort, wo Menschen zusammenkommen, muß die Architektur zurücktreten, zum Hintergrund werden, verschwinden. Aber es muß eine Illusion bleiben.« Deshalb sollte eine solche Hülle nicht zu bestimmt sein; man müßte sie erweitern können mehr durch Andeutungen als in der Realität. Der in Erscheinung tretende Raum müßte ein Gefühl der Entspannung vermitteln; darin gedeiht die Geselligkeit.

Dann aber spricht er von der Forderung nach Intimität inmitten der Geselligkeit. Man muß sich zurückziehen können. Segal ermahnte die Wohnbauarchitekten, diese Bedürfnisse potentieller Bauherren sorgfältig zu studieren. Wenn man keinen Platz vorsieht für ein Ausweichen, sei es zur Ruhe, zum Fernsehen, zum Nach-

The photograph from which Arthur Segal made his painting of »La Casa Piccola«.
Nach diesem Foto malte Arthur Segal das Bild »La Casa Piccola«.

Lewisham self-built houses, street view (see p. 175–179)
Selbstbauhäuser in Lewisham, Straßenseite (vgl. S. 175–179)

not create space for gregariousness. Many years after publishing his housing studies, Segal was distressed that: »Housing architects have hardly taken notice, even now, of television and transistor. They talk about sound insulation; but at the same time they are busily engaged in destroying the differentiated house plan and remain obsessed with the old-fashioned concept of unifying space in houses.«

»As I see it, buildings are there to be a background for people, against which they move, a background which envelopes them, protects them, gives them pleasure, and allows them to add a little bit of themselves. Let me describe something which I would say describes architecture. Imagine a very narrow street in the South with sun, high walls or buildings on either side. You see a woman coming towards you. That woman dominates the scene. The street is hers, the road is there to serve her purpose, namely to go from one point to another. It does not overwhelm her with its own needs. We should always understand that the buildings we are putting up are not made of living material, they are only there to serve the living.«

»Rote Äpfel«. Dynamic equilibrium in space as seen by Arthur Segal. Painting of 1935 (see p. 62).
»Rote Äpfel«. Dynamisches Gleichgewicht im Raum, wie es Arthur Segal sah. Gemälde 1935 (vgl. S. 63).

denken oder für die Musik, kann man keinen Platz für Geselligkeit schaffen. Viele Jahre, nachdem er seine Wohnbaustudien veröffentlicht hatte, stellte Segal bekümmert fest, daß »Wohnbauarchitekten bis heute Fernsehen und Transistorradios kaum zur Kenntnis genommen haben. Sie reden über Isolierung und zerstören gleichzeitig den differenzierten Grundriß in ihrer Besessenheit für das altmodische Konzept des alle vereinenden Raumes.«

»So wie ich es sehe, müssen Häuser den Hintergrund für die Menschen bilden, die sich davor bewegen, ein Hintergrund, der sie umhüllt, beschützt, ihnen Freude macht und ihnen die Möglichkeit gibt, ein klein wenig von sich selbst hinzuzufügen. Laßt mich etwas beschreiben, von dem ich meine, das es die Architektur beschreibt. Stellt Euch eine ganz enge Straße vor, im Süden, Sonne, hohe Hauswände an beiden Seiten. Ihr seht eine Frau auf Euch zukommen. Sie beherrscht die Szene. Die Straße gehört ihr. Die Straße dient ihr. Die Bedürfnisse der Straße dürfen sie nicht überwältigen. Wir sollten niemals vergessen, daß die Gebäude, die wir errichten, nicht aus leben-

In his house planning, Segal looked for balance. He used the word repeatedly, and it can be seen in his plans. Shying from pattern-making and from generalising, he thought of groups, individual families and individuals, and always sought a balanced optimum rather than a panacea. This balance is clear at the scale of the residential community in the studies of indigenous architecture of the Baleares, in his layouts in *Home and Environment*, and right up to »Walter's Way« in Lewisham. It is also clear within the dwelling itself, and most obviously in the »Mediterranean« archetype he studied in the Baleares, where the central hall, the *sala*, is surrounded by smaller »houses« for sleeping and for service – a gregarious arrangement miniaturised in his own »little house«.

The centre of the dwelling is not anymore symbolically directed, it is formalised neither in front of the great fireplace of old, nor round that overstimulating space invader of the second half of our century, the television. But it is a focus for action, whose boundaries dissolve and blur as the action focuses attention. In an early project, Segal proposed a circular living room. It was too diagrammatic a design, but fascinating for its investigating a living room which, »contrary to usual practice, has no focal point. The idea being that people would sit round in a circle and be themselves, so to speak, the focus.«

»If architecture is to reach the mind – beyond the fleeting visual impressions – there must come into action a meaningful background against which all this action can take place. This happens when a building can display *a fuller understanding of its users* than is achieved by mere servicability; and when, in turn, *it is likewise understood by them*; in such reciprocal rapport a level of relationship is produced which is conducive to mutual well-being....«

One element of conviviality is the essential relationship between architect and client. The traditional split – the notion that architecture has become a profession within George Bernard Shaw's definition as »a conspiracy against the laity« – was something Segal felt instinctively. Indeed he never trusted architects, and was known to have advised against the use of one! »I am not an architect!« he even said.[2]

The architectural profession, in its obsession with form-making, lost its central role as catalyst for the convivial provision of building. And here the final ingredient is making a building *process* which itself is convivial.

Segal's aim, quoted in the concluding words of his address to the RIBA in May 1977, was that, »We could not only bring back the joy of building but thereby make people happier. But this can only be done if we succeed in involving the public to a much greater degree. If this happens, it will give us, as architects, a new identity. It will also give us a different status, because we will no longer be people that are on the defensive. Then we will be able to offer much more than now. There is a large amount of design ability and design desire in the public, and this must be released. And by releasing it we shall get results that will help us out of our present impasse.«

Segal's work has touched a deeper chord than those encompassed by ideas of »simple honest building« or of »decent housing provision,« admirable as each of these is. In many other times, with less specialised roles and with that ancient communality, the whole village would »make« the new house required; we have images like the 19th century »barn-raising« as celebration of creative communality.

Segal's achievement is to make that easy, to facilitate that again for our fragmented, individualised world. The little groups of separate Segal houses on the hillsides in south London have a stronger cohesive agent than street form, building line or cornice level: they are the homes of friends, of friendships made through shared activity. The architecture embodies the shared values as they act as catalyst of this social cohesion. Segal's interest had become the means and process of housing and of inhabitation, not the production of objects. All the Lewisham self-builders, as we have seen, were chosen by lottery from the large number of ordinary working-class applicants from the waiting list for public housing. When the first group made a video film of the whole thing, based on interviews with them all, a leading local socialist politician suggested that they were all becoming prima donnas. »Nonsense!« they replied. »But,« they added, »now we'll try *anything*; there's nothing we won't try!«

Segal's achievement inhabits a different world from one view of what is currently called »community architecture«. There is none of the benevolent paternalism with which Britain's best-known »community architect« is seen on television stopping by to offer an old lady advice about her blocked rain gutter. That image, the local lad who's become an expert and can offer us a special skill in time of need, is typified in Britain's love of James Herriot's archetypal vet.

dem Material sind, sie sind nur da, um den Lebenden zu dienen.«

In seinen Grundrissen strebte Segal nach »Gleichgewicht«. Er brauchte das Wort dauernd, und man sieht es in seinen Plänen. Er scheute zurück vor Mustern und Verallgemeinerungen. Er dachte an Gruppen, Familien und Individuen, und suchte immer nach einem ausgewogenen Optimum. Diese Ausgewogenheit wird klar in seinen Studien der bodenständigen Architektur auf den Balearen, in seinen Situationsplänen in *Home and Environment*, bis zum *Walters' Way* in Lewisham. Sie wird auch innerhalb der Wohnungen klar, ganz besonders in denen nach dem mediterranen Typus mit der zentralen Halle und kleinen »Häusern« zum Schlafen und Wirtschaften darum herum: Ein geselliges Arrangement, dessen Miniaturausgabe sein *kleines Haus* ist. Das Zentrum der Wohnung ist nicht mehr symbolisch vor dem großen Kamin wie in alten Zeiten, auch nicht um den ewig stimulierenden Eindringling der zweiten Hälfte unseres Jahrhunderts, das Fernsehgerät, es ist als Zentrum von Aktionen gedacht. Es ist der Ort, wo alle hinsehen. In einem frühen Projekt schlug Segal einen runden Wohnraum vor: ein Raum, »der entgegen der üblichen Praxis keine Blickrichtung hat. Die Idee war, daß die Leute in einem Kreis sitzen und selbst gewissermaßen den Blickpunkt bilden.«

»Wenn Architektur den Geist ansprechen soll – über die fließenden visuellen Eindrücke hinaus – muß es eine tieferreichende Basis geben, vor der sich alles abspielt. Das geschieht, wenn ein Gebäude ein breiteres Verständnis für seine Benutzer ausdrücken kann, über die einfache Nützlichkeit hinaus und wenn es dann auch von den Benutzern verstanden wird; in solch einer Wechselwirkung entsteht eine Beziehung, die zum allgemeinen Wohlbefinden führen kann ...«

Die traditionelle Spaltung zwischen Architekt und Bauherrn ist nach einer Definition von Bernard Shaw »ein Komplott gegen die Laien« geworden. Das spürte Segal instinktiv. Er traute niemals einem Architekten, und man weiß, daß er davon abgeraten hat, solche zuzuziehen. »Ich bin kein Architekt«, sagte er sogar. (Der Segal-Trust lehnt es deshalb auch prinzipiell ab, einen Architekten zu empfehlen.) Die Architekten, in ihrer Besessenheit für die Form, haben ihre Rolle als Katalysatoren beim gemeinsamen Schaffen eines Gebäudes verloren. Ihre Aufgabe wäre, zu einem gemeinsamen Bauprozeß zu verhelfen.

Segal formulierte sein Anliegen in den Schlußworten eines Vortrages vor dem Royal Institute of British Architects im Mai 1977: »Wir müssen die Öffentlichkeit in weit höherem Maße teilhaben lassen. Wenn wir das erreichen, wird es uns, als Architekten, eine neue Identität geben und auch einen anderen Status, weil wir nicht länger Menschen in der Defensive sein werden. Dann werden wir in der Lage sein, viel mehr zu bieten als heute. Es gibt in breiten Schichten die Fähigkeiten und die Bereitschaft zu planen, und wenn man die freisetzt, wird man Ergebnisse erzielen, die uns aus unserer gegenwärtigen Sackgasse heraushelfen.«

In anderen Zeiten, mit weniger eng umschriebenen Rollen und einem stärkeren Gemeinsinn, wäre das ganze Dorf daran gegangen, das Haus, das man brauchte, »zu machen«; wir kennen Beispiele aus dem 19. Jahrhundert, als das Aufrichten einer Scheune ein Fest kreativer Gemeinsamkeit war. Segals Verdienst ist es, solche Vorgänge in unserer zerstückelten, individualisierten Welt wieder möglich gemacht zu haben. Die kleinen Gruppen einzelstehender Segal-Häuser an einem Hang in Süd-London haben einen starken gemeinsamen Nenner: Sie sind die Häuser von Freunden. Die Freundschaft wuchs aus gemeinsamem Schaffen. Die Architektur bringt diese gemeinsamen Werte zum Ausdruck, war Katalysator des sozialen Zusammenhalts.

Segals Interessen galten den Prozessen des Behausens und des Bewohnens, nicht der Produktion von Objekten. Diese Selbstbauer in Lewisham waren, wie wir gesehen haben, durch das Los aus der langen Warteliste der Wohnungsbehörde ausgewählt worden. Größtenteils waren es Arbeiter. Als die erste Gruppe einen Video-Film des Bauvorgangs aufnahm, mit Interviews der Beteiligten, meinte ein führender sozialistischer Politiker, daß sie nun alle *Primadonnen* würden. »Unsinn«, antworteten sie, »aber nun werden wir alles versuchen, es gibt nichts, das wir nicht in Angriff nehmen werden.«

Das ist eine andere Welt als das, was man heute gemeinhin als »Architektur der Gemeinschaft« bezeichnet. Darin ist nichts von diesem wohlwollenden Paternalismus, mit dem Englands bekanntester Architekt auf diesem Gebiet auf dem Bildschirm zu sehen ist, wie er stehenbleibt und einer alten Dame Rat für ihre verstopfte Dachrinne gibt.

Segal und Jon Broome in Lewisham hatten eine andere Aufgabe; sie ermöglichten gewöhnlichen,

Self-builders of different generations, helping each other at Lewisham.
Selbstbauer verschiedener Generationen helfen einander gegenseitig (Lewisham).

Groups of self-builders at Lewisham – with architect.
Selbstbau-Gruppen in Lewisham – mit dem Architekten.

Segal and Jon Broome in Lewisham had quite a different role; enabling ordinary unskilled people, severally and jointly, to regain their own powers of dwelling. (And in this it is equally far from typical self-build organisations.)

But Eupalinos was no more likely literally to build with stones himself than was Segal with wood. Certainly no master craftsman himself, Segal made heavy weather of putting up a shelf. We don't need to share all our roles to create convivial building.

He was, however, deeply concerned about the traditionally anti-social and hazardous nature of the building industry and this helped form his own method of construction. »I am thinking about the need to improve the working conditions in the industry which will allow it to attract more capable men and women. We offer in the building industry some of the worst working conditions that exist.« He was equally scathing of the nostalgia for outdated craftsmanship and glorification of »honest toil« which he felt had trickled right through to Gropius from William Morris. »Would Morris have liked the look of a Black & Decker drill?« he asked. »I fear not.« All Segal's activity has implicitly followed that motto which is, however, attributed to Morris: »I want to design things that people get pleasure in making, and to make things that people get pleasure in using.«

»Over the gable of a fine old timber house in the Bernese Oberland, 200 years old, are written the words ›Master K. and his men did everything to my entire satisfaction.‹ Please note,« added Walter Segal, »not merely to his ›reasonable‹ satisfaction as stipulated in the RIBA Contract. On how many houses in post-war England could this be written?«[3]

So this continuum of convivial activity deserves celebration.

»Eventually the roof was finished. This occasion obviously called for some sort of celebration in keeping with traditional customs, and so Don Antonio asked me to arrange with my client for the usual big meal for all the people connected with the job. It was to be taken at the *fonda* (restaurant) «Las Palmeras», behind the colonnades surrounding a bright flower, fruit and vegetable market.

»There we all met one Saturday at noon, and sat around a big table, all 15 or 20 of us, with the client, the builder, and the architect. Miguel (the foreman) was host, and I must confess I have never eaten as much in all my life as on that day.

»First of all we had a big dish of excellent rice, cooked with all sorts of vegetables and contributions from all available animals on the land and in the sea, from mutton to mussels. It was served from an enormous pan, which was so laden that in the beginning it hid Miguel, in front of whom it was placed, almost completely.

»I must say, however, that he soon reappeared behind the rice mountain, which was rapidly losing height. We used deep plates, which were filled up, and all the others had a second helping, while my refusal made me suspect of not quite having enjoyed the dish. There was white and red wine, but nobody drank a lot, most people diluting it with water as is the habit in the South since the days of Homer.

»It was a Saturday, and nobody was worried. The *mujeres* (women) had been told, and the next bull-fight was still a week ahead. So we had four hours of good time until everything was eaten, all the stories told, and everybody was ready for a good rest....

»There is little to add. I had enjoyed more than ever before being an architect.«

Notes and Quotations

1. Segal was talking just after Le Corbusier's death; this is from his most thoughtful and provocative obituary in *The Architects' Journal* in September 1965.
2. In this tradition, the Walter Segal Trust finds the last thing it is inclined to do is recommend an architect!
3. Anonymous leader in *Architectural Design*, November 1953, p299; see note 3, p128.

Quotations from Walter Segal in this section from: 1965(B); 1973(B)3; 1977(B)2; 1945(B)3; Wills & Yeo interview 1984; 1953(B); 1942(B).

nicht ausgebildeten Leuten, einzeln und gemeinsam, ihre Fähigkeit, die eigene Behausung zu schaffen, zurückzugewinnen. Und in dieser Hinsicht grenzen sie sich auch von den üblichen Selbstbau-Organisationen ab.

Aber Eupalinos hätte wahrscheinlich ebenso wenig selbst, wörtlich genommen, die Steine aufeinander geschichtet, wie Segal mit Holz gearbeitet hätte. Handwerklich war er selber kein Meister. Es war für ihn ein schwieriges Unterfangen, ein Regal aufzustellen. Wir brauchen nicht selbst alle Rollen zu spielen, um einen gemeinsamen Bau zu schaffen.

Er war jedoch tief betroffen von den traditionell antisozialen und unfallträchtigen Arbeitsbedingungen in der Bauindustrie. »Ich denke an die Notwendigkeit, die Arbeitsbedingungen in der Bauindustrie zu verbessern, damit sie wieder fähigere Frauen und Männer anziehen kann. Hier werden heute die schlechtesten Arbeitsbedingungen geboten, die es gibt ...«

Auch kannte er nur beißende Kritik für veraltete Handwerksmethoden und die Glorifizierung der ehrlichen Arbeit, etwas, das, wie er meinte, von William Morris geradewegs zu Gropius durchgesickert war. »Würde ein Black&Decker-Bohrer William Morris gefallen?« fragte er. »Ich fürchte, nicht!«

Alle seine Aktivitäten standen jedoch unter einem Motto, das Morris zugeschrieben wird: Ich möchte Dinge entwerfen, die man gerne herstellt, und Dinge herstellen, die man gerne benutzt.

Über dem Giebel eines schönen, 200 Jahre alten Holzhauses im Berner Oberland stehen die Worte: Meister K und seine Männer taten alles zu meiner vollen Zufriedenheit. »Bitte zu bemerken«, fügte Walter Segal hinzu, »nicht nur zu einer angemessenen Zufriedenheit, wie es in den RIBA-Verträgen festgelegt ist. An wie viele Häuser im Nachkriegs-England könnte man das heute schreiben?«

»Einmal war dann das Dach fertig, und dieses Ereignis verlangte offensichtlich, nach altem Brauch, eine Feier, und so bat mich Don Antonio, mit dem Bauherrn das übliche große Mahl für jeden, der am Bau beteiligt war, zu arrangieren. Es wurde in der *fonda* (Restaurant) »*Las Palmeras*« eingenommen, hinter den Kolonnaden, die den leuchtend bunten Blumen-, Gemüse- und Obstmarkt umgeben.

Dort trafen sich alle am Samstag mittag und saßen um den großen Tisch, 15 oder 20 Leute, mit dem Bauherrn, dem Baumeister und dem Architekten. Miguel, der Vorarbeiter, war Gastgeber, und ich muß gestehen, daß ich niemals in meinem Leben so viel gegessen habe wie an jenem Tage.

Zuerst hatten wir eine große Platte mit ausgezeichnetem Reis, gekocht mit allen Arten von Gemüse und Beigaben aller greifbaren Tiere, zu Lande und zur See, vom Hammel bis zur Muschel. Es wurde in einer riesigen Pfanne serviert, die anfangs so voll geladen war, daß Miguel ganz dahinter verschwand.

Ich muß jedoch sagen, daß er bald hinter diesem Reisberg, der ständig an Höhe verlor, wieder hervorkam. Wir aßen aus tiefen Tellern, die ganz voll gefüllt wurden, und alle andern nahmen zum zweiten Mal, und meine Weigerung führte zu dem Verdacht, daß ich das Gericht nicht wirklich gemocht hätte. Es gab weißen und roten Wein; aber niemand trank viel, die meisten verdünnten ihn mit Wasser, wie es im Süden der Brauch ist, seit den Zeiten Homers.

Es war Samstag, und niemand war in Eile. Den Frauen hatte man Bescheid gesagt, und der nächste Stierkampf fand erst in einer Woche statt. So hatten wir gute vier Stunden Zeit, bis alles gegessen, alle Geschichten erzählt, und jedermann für eine lange Ruhe bereit war ...

Es bleibt wenig hinzuzufügen. Ich freute mich mehr als je zuvor, ein Architekt zu sein.«

Finally

»It is not necessary to know whether our existence is significant, but the fight for insight is an act of asserting freedom in which we can rejoice.«

That fine phrase was how Walter Segal ended his thoughtful analysis of Fritz Schumacher's *Small is Beautiful* which I had sent him to review. He was rejecting the religious underpinning of Schumacher, just as positively as he had rejected dogmatic Marxism in his youth. And he was asserting what he called his »immensely strong animal confidence«.

He was described by one of his AA students in the 1940s (who later, as an editor, became the instigator of much of his writing), as: »an absolutely stunning combination of an immensely practical architect and an intellectual, with prodigious energy and application.«[1]

He drove himself very hard, and he gave himself space to enjoy a life much fuller than just architecture. His intense concentration and short attention span were among the many childlike qualities he retained. »He is one of today's liveliest young architects,« I wrote when he was nearly seventy. But he probably never looked youthful, in the way for example of Americans, obsessed with fitness and physical perfection. Indeed his

Lewisham: Human background for living.
Lewisham: Lebensräume.

Zum Schluß

»Es ist nicht nötig zu wissen, ob unsere Existenz Bedeutung hat; aber der Kampf um Einsicht ist ein Akt, sich Freiheit zu erkämpfen. Darin können wir freudig leben.«

Mit diesem schönen Satz beendete Walter Segal seine nachdenkliche Analyse von Fritz Schumachers *Small is Beautiful*, dem Buch, das ich ihm zur Besprechung geschickt hatte. Er lehnte den religiösen Unterbau bei Schumacher ab, ebenso bestimmt wie seinerzeit den dogmatischeren Marxismus seiner Jugend. Und er verteidigte, was er sein »ungeheuer starkes animalisches Vertrauen« nannte.

Einer seiner Studenten aus den Vierziger Jahren bei der Architectural Association (der später als Redakteur ihn zu vielen seiner Artikel angeregt hat), beschrieb ihn als »eine absolut verblüffende Kombination des ungeheuer praktischen Architekten und des Intellektuellen mit einer Fülle an Energie und Hingabe«. (Colin Boyne, in seinem Dank an Walter Segal nach seinem RIBA-Vortrag, Mai 1977)

Er arbeitete sehr hart, und er bewahrte sich Raum, ein Leben zu genießen, in dem es anderes gab als nur die Architektur.

Seine intensive Konzentration und sein kurzes Aufmerken waren einige der vielen kindlichen Eigenschaften, die er beibehielt. »Er gehört zu den lebhaftesten jungen Architekten«, habe ich geschrieben, als er fast siebzig war. Aber wahrscheinlich hat er niemals jugendlich ausgesehen, in der Art wie etwa die Amerikaner, die besessen sind von Fitneß und physischer Perfektion. Tatsächlich war der Eindruck seiner ersten Assistenten, daß »er viel älter schien, voller Erfahrung«. (Roger Walters, Notiz an den Autor, April 1988) Seine gnomenhafte Gestalt, der rutschende Gürtel, ein alter Pullover (man sah ihn niemals in Jacke oder Mantel, wie immer auch das Wetter war) und seine Kinderschuhe, alles das zeigt eine ganz andere Auffassung von Jugendlichkeit.

Segal wandte wenig Zeit auf für die Alten im Geist. Seine Generation war über mehr als die Hälfte seines Lebens nicht die seine gewesen. »Wenn Menschen einmal 50 sind, finde ich sie schon nach ein paar Minuten nicht mehr wirklich anregend oder amüsant; sie haben natürlich ihre eigenen Probleme, ich verstehe das; aber es sind nicht die meinen.«

»Ich habe ein Verlangen nach neuen Stimuli und Eindrücken. Man wird so leicht altbacken«, fügte er hinzu, während seine Gedanken weiter wanderten zur Renovierung alter Gebäude. »Alte, neu aufgeputzte Gebäude sind eine miserable Sache. Ich denke, die tote Hand der Vergangenheit sollte bleiben, was sie ist, nämlich tot.« Er haßte es, in alten Häusern zu leben. In einem Haus aus zweiter Hand zu wohnen, war für ihn wie das Tragen fremder Wäsche, sagte er. (Neville Conder, Brief an den Autor, 9. 5. 88) Ich zog zweimal um während der Jahre, die ich ihn kannte, jedes Mal in ein altes Haus; er wurde niemals müde, all das Unheil zu beschreiben, das mit der Zeit zutage treten würde, mit dem Verfall dieser unzulänglichen Gebäude.

Segals Jugendlichkeit, die über Jahre ihm die Wertschätzung der Studenten brachte, kam auch in seiner lebhaften Heiterkeit und seiner mitreißenden Begeisterung zum Ausdruck, in den nicht seltenen kindischen Streichen »im Mozartgeist«, die er mit einem Zwinkern im Auge erzählte oder dem plötzlichen, sehr unmusikalischen Entgleiten in ein altes Tessiner Volkslied.

Kindlichkeit ist nur ein Aspekt der Gabe, Dinge zu durchschauen, sich selbst zu vergessen, oder des Kaisers neue Kleider als das zu sehen, was sie nicht sind. »Der Mensch ist ein denkendes Rohr im Winde.« Aber seine großen Werke vollbringt er dann, wenn das Denken nicht vorgesehen war. »Kindlichkeit muß wieder hergestellt werden in langen Jahren des Trainings, sich selbst zu vergessen«, sagt der Zen-Meister D.T. Suzuki (im Vorwort zu Zen und die Kunst des Bogenschießens). Ein anderer Zen-Meister erklärt, daß Lao-Tse, dessen Name mit »alter Mann« übersetzt wird, besser wiederzugeben wäre mit »alter Jüngling«.

very first assistant's first impressions were that »he seemed much older and full of experience.«[2] The pixie form, slipping belt and old jersey (he was never seen in jacket or coat whatever the weather), and his Startrite (child's) shoes, all imply a rather different glimpse of youth.

Segal had little time for the elderly in spirit, and the generation born with him had not been his generation for half his life. »When people get to be over fifty, after a few minutes contact I find them rarely stimulating or amusing; they tend to have their own problems and, of course, I understand this. But they are not my problems. I've got quite a craving for new stimuli and impressions.

»One gets stale so easily,« he added, as his train of though moved on to the rehabitation of old buildings. »Old buildings spruced up are a miserable thing; I think the dead hand of the past should be completely content to remain just that: dead.« He hated living in an old house; to live in a secondhand house was like wearing somebody else's underwear, he said.[3] (I moved twice during the years I knew him, from one built for myself to old houses; he never ceased listing all the disasters which would slowly reveal themselves as the inadequate old building decayed.)

Segal's youth, which over the years endeared him to students and they to him, was also in the bustling glee, infectious enthusiasms, and a not too rare Mozartian childish prank recounted with a glint in the eye, or a sudden most unmusical lapse into an old Ticinese folksong.

Childlike is but one aspect of the discipline of »seeing through«, of self-forgetfulness, of recognising the emperor's new clothes for what they are not. »Man is a thinking reed, but his great works are done when he is not calculating or thinking. »Childlikeness« has to be restored with long years of training in self-forgetfulness,« said the Zen master D T Suzuki[4]. (Another Zen master explains that Lao-Tse, whose name translates as »old man«, is more precisely rendered as »old youngster.«)

»He never seemed to age. He spoke of the autumn of his life but it was an autumn which never made one think winter a possibility....« Twenty years after Segal wrote this (in his obituary of le Corbusier), he described Mendelsohn as »one of the outstanding architects of the twentieth century, with Wright, le Corbusier, Mies and Gropius«. When I expressed slight surprise at such an opening (wasn't it just a little over the top?), he replied without hesitation: »But John, I didn't promise to tell the truth!«

»So spake he with twinkling eyes and twisted brows, glancing hither and thither with long-drawn whistling breath....«[5]

As he ended the review of Bruno Zevi's book on Mendelsohn, which was almost the last piece he wrote, Segal's optimism was certainly heartfelt. Quoting Zevi on »the dangers of a regression to the pre-modern and neo-conservative« in the mid-1980s, Segal commented: »He need not despair, for the post-modern spook is on the way out. A few more years until we build again.«

When he wrote that Walter Segal knew that he had not long to live. Having spent a vigorous, active life claiming never to have consulted doctor, in his later 70s he was finally admitted urgently to hospital. Amazingly soon he was up looking fitter than for years, but he knew it was a reprieve. Now the second Lewisham project, perhaps his most characteristic work, was launched and well on its way.

In his last weeks, although his intellectual energy was as infections and active as ever, he privately felt his battles were fought. One Sunday at the end of October 1985 he fulminated over tea with his guests about community architecture, about housing not being »architecture«, and all manner of things as always. His friends left and his wife went out for a while. He died, unpredictably, at home before supper.

»Farewell, dispenser of favours, guide, giver of good things«
(end of the Second Homeric Hymn to Hermes)

Notes and Quotations

1. Colin Boyne, offering vote of thanks to Walter Segal after his RIBA lecture, May 1977.
2. Roger Walters, note to the author, April 1988.
3. quoted by Neville Conder, letter to the author, 9 May 1988.
4. in his forward to *Zen and the Art of Archery*.
5. Hermes, in the *First Homeric Hymn to Hermes*, quoted from Charles Boer's translation, Spring Publications, 1970.

Quotations from Walter Segal in this section from: 1974(B)1; 1976(C)1; 1965(B); review in *AJ* 22 May 1985.

»Er schien niemals zu altern. Er sprach vom Herbst seines Lebens; aber er war ein Herbst, der den Gedanken an einen Winter nicht aufkommen ließ.«

Zwanzig Jahre nachdem Segal das geschrieben hatte (in einem Nachruf für Le Corbusier), beschrieb er Mendelsohn »als einen herausragenden Architekten des 20. Jahrhunderts, zusammen mit Wright, Le Corbusier, Mies und Gropius ...« Als ich eine leichte Überraschung über diesen Anfang ausdrückte (war das nicht ein bißchen zu hoch gegriffen?) antwortete er ohne zu zögern: »Aber John, ich habe doch nicht versprochen, die Wahrheit zu sagen.« »Also sprach er und zog mit blinzelnden Blicken die Brauen gar gewaltig empor und schaute hierhin und dorthin, munter pfiff er ...« (Hermes in der zweiten Homerischen Hymne, deutsch: Thassilo von Scheffer, 1927).

Am Ende seiner Besprechung von Bruno Zevis Buch, einer der letzten, die er geschrieben hat, kam Segals Optimismus gewiß von Herzen. Er zitierte Zevi zu »den Gefahren eines Rückfalls in die Pre-Moderne oder Neo-Konservative« in der Mitte der Achtziger Jahre und bemerkte: »Er braucht nicht zu verzagen; denn der post-moderne Spuk ist im Abflauen. Noch ein paar Jahre, und wir werden wieder bauen.«

Als Walter Segal das schrieb, wußte er, daß er nicht mehr lange zu leben hatte. Nach einem tätigen, aktiven Leben, in dem er niemals einen Arzt konsultiert hatte, wurde er in seinen späten Siebzigern plötzlich als Notfall ins Spital eingeliefert. Er war erstaunlich schnell wieder auf den Beinen, sah besser aus als seit Jahren; aber er wußte, es war nur ein Aufschub. Jetzt war das Lewisham-Projekt, vielleicht sein charakteristischstes Werk, angelaufen und auf gutem Wege.

Während seiner letzten Wochen, obwohl seine intellektuelle Energie so aktiv und ansteckend wie immer war, fühlte er, daß seine Schlacht geschlagen war. Eines Sonntags, Ende Oktober 1985, ereiferte er sich beim Tee mit seinen Gästen über Gemeinschafts-Architektur, darüber, daß Wohnbau keine Architektur sei und viele andere Dinge, wie immer. Seine Freunde gingen, und auch seine Frau ging kurz hinaus. Er starb, unvorhergesehen, zu Hause, vor dem Abendessen.

»Und so sei mir gegrüßt ... Du meines Liedes Beginn ... Heil Dir, Freudenspender. Geleiter, Geber des Guten.« (Ende der zweiten Homerischen Hymne an Hermes)

Lewisham: Quiet precision.
Lewisham: Die Ruhe der Präzision.

Appendix A / Anhang A
Buildings and Projects 1932–1985
Bauten und Projekte 1932–1985

This list of Walter Segal's major buildings and projects is based on that published in *The Architects' Journal*, 4 May 1988. That list was taken from an exhaustively researched list (which I have not seen) by Philip Christou, who performed the invaluable task of Segal archivist in 1986-7.
[P] indicates a project; [D] a building demolished or heavily altered. Dates indicate the start of each work.

1929
Small house competition, (with KH Beyerling & A Bloch) published entry, (commended) [P].

1932
Casa Piccola, Ascona
additions and interiors and »other tiny projects« in Ascona for 12 months.

1933
House, »son anmadones«, Palma de Mallorca, Spain
Terrerno, Palma
Calle Dos De Mayo, Palma
another house, Mallorca
restaurant, Ibiza Ciudad Marina,
Verner y Gertrudis, Ibiza.

1937
News Chronicle schools competition (commended)
Factory for Premier Pickle Co, 37, Holly Street, London E8 [D].

1940–4
Ministry of Supply hostels for workers (with Bertram Carter)
at: Blackpole, Gloucestershire (for 500 people); Steeton, Gloucestershire (for 500 people); Kirby, Liverpool (two hostels, each for 1000 people).

1946
Camp School for 30 resident boys, Hawspur (project).

1947
Schultz House, corner Lawrence Road/The Ridgeway, Barnet, London NW7.

1948
16 flats, 149, Leigham Court Road, Streatham, Lambeth, London SW16.

1949
ABSE prototype aluminium frame house [P].

1950
8 houses, 1–8, St Anne's Close, London N6 (assistant Alan Blanc)

6 houses, West Heath Gardens, Barnet, London NW3 (assistant Alan Blanc)
Lennard House, 20 Highview Gardens, Barnet, London N3 (assistant Alan Blanc)
12 flats, 7&9, Chesham Street, London SW1.

1952
Weil House, 16 & 18 Neeld Crescent, Barnet, London NW4.

1953
2 houses, 2 & 4 Woodville Road, Brent, London NW6
Sanders House, 18 Colney Hatch Lane, London N10 [D]
Niven House, Dennis Lane, Harrow, Middlesex [D].

1954
Offices for Tretol Ltd (first phase), corner Edgeware Road/ Holmstall Avenue, Colindale, Brent, London NW9
Offices and factory The Premier Pickle Factory, Ramsgate Street and Tyssen Street, Hackney, London E8 [D]
Block of 14 flats, 43 & 44 Rutland Gate, London SW7.

1955
Block of 12 flats and maisonettes, 22–26 Ovington Square, Kensington and Chelsea, London SW3
Twin Block Sky Gardens, or Twin Block Hanging Gardens, Hemel Hempstead New Town, Hertfordshire [P].

1957
Ski House, near the Skihaus Arflina, Fideris, Graubünden, Switzerland
B. Rowley House, 15 The Boltons, London SW 10 [D]
Weiss House, Rear of Tunstall House, Royal Avenue, Worcester Park, Epsom and Ewell, Surrey [P].

1958
A. H. Rowley House, site adjacent to 432 Church Street, Enfield, London N9.

1959
Tobler House, Rugby Road, Richmond, London [D].

1960
Block of four maisonettes and four garages, 51a Compayne Gardens, London NW6
Twin block of 12 flats, corner of Brookhill Road and Cat Hill, Barnet, London
Block of 14 flats, Kingsmere Court, Salmon Street, Brent, London NW9
Block of six flats and four garages, Bedfont Road, Hounslow, London.

1961
Gutt House, 79 West Heat Road, Barnet, London NW3.

1962
Architect's own house, 9 North Hill, Haringey, London N6.

1963
Temporary house at rear of garden of architect's own house, 9 North Hill, Haringey, London N6
Terrace of three houses, 4, 5 and 5a Tasker Road, Camden, London NW3
Tretol Offices phase 2 (addition of new south and west wing and third floor added to existing building), corner of Edgware Road and Holmstall Avenue, Brent, London NW9
Local authority terrace of houses, 36 & 38 Gordon House Road, Camden, London NW5 [P]
›Twin Block Sky Gardens‹, site fronting Jonathan Street, Lambeth, London SE11 [P]
Sisalcraft Pavilion, IBSAC Exhibition, Crystal Palace, London
Donohue House, Ballaghkean, North Wexford, Ireland.

1969
Leigh House, Main Street, Yelling, Huntingdon, Cambs.
Collier House, ›Tree House‹, 30 Chapel Street, Halstead, Essex
Vesey Holt House extension, ›Phantom Ranch‹, North Common (near Lewes), East Sussex.

1970
Rietty Playroom, 40 Old Church Lane, Brent, London NW9.

1971
Lomask House, Ballycummisk, County Cork, Ireland

Cook House, Warrenorth, North Common (near Chailey), East Sussex
Holland House, School Lane, Bromeswell (near Woodbridge), Suffolk.

1972
Children's Home, Singleton, West Sussex
Godfrey House and Surgery, Chapel Lane, Clifford, Boston Spa, Yorkshire (assistant Tony Cooper).

1975
Birch House, ›Telfs‹, Hendon WoodLane, Barnet, London NW7.

1976
Houses for Solon Housing Association, 18–24d Albermarle Road, Bormley, Kent [P]
11 single storey houses for Toothill Self-build Housing Association, Thamesdown [P].

1977
11 single storey, three double-storey self-build houses, Lewisham (with Jon Broome), Segal Close, SE23. 11 & 13 Elstree Hill. 11 Longton Avenue, corner of Longton Avenue and Ormanton Road, SE26. 17a & b Longton Avenue, SE26.

1979
Green House, Kennell Hill, Sharnbrook, Bedfordshire.

1984
Hackett Barn, Hillcross Stockland (near Honiton), Devon.

1985
13 double-storey self-build houses (with Jon Broome), Walter's Way, Honor Oak Park, Lewisham, London SE23.

Selected Writings by Walter Segal
Schriften von Walter Segal

Segal wrote extensively throughout his life; there are a large bunch of housing and layout studies, later of book reviews, and then of short (anonymous) gossip notes. There are a few major lectures, analytical studies and contributions to books. This first short select list includes only Segal's own written work.

1934
»L'Architecture de l'ile d' Ibiza« (with photographs by Raoul Hausmann) *Oeuvres*, No 7, September 1934, pp15–18. A longer version, »The Domestic Architecture of Ibiza« by Walter Segal (translated by Daphne Sanders), remains unpublished.

1935
»Bijut el Fellahin: habitations des paysans en Egypte« *Oeuvres*, No 10, 1935, pp13–21. Written and illustrated in Cairo, 20.9.1935 in French.
Thesis on Egyptian Dynastic Chairs; not completed. Voluminous MSS material deposited with Dr Marianne Eaton-Krauss, Munster.

1936
»L'Habitation Rurale a Mallorca« *Chantiers*, 10.36, pp521–525 and again in *Oeuvres*, No 37, 1936, pp8–11.

1939
»Bruno Taut« (Obituary) *Journal of the Royal Institute of British Architects* (RIBAJ), January 1939, p313 (first English publication).

1942
»Building a house in Majorca« *The Architect and Building News* (A&BN), V170, 19 June 1942, p160–2.

1945
Planning and Transport: Their Effects on Industry and Residence, J M Dent & Sons, Ltd. (for the Cooperative Permanent Building Society), London 1945. One of eight books in a series called »Building and Society«.
»Up Your Street; The Neighbourhood in Relation to Housing« Article by WS about his »Up your street« exhibition (assisted by Philip Powell and Patricia Owen) *The Architects' Journal* (AJ), 1945, pp97–100
»Design Sketching« *Building*, September 1945, pp 238–241
Germanomania (lengthy, unpublished thesis on Nazism) date unknown.

1946
»Building now« *Building*, May 1946, pp152–5
Homes for the People: How Modern building technique can provide high standard dwellings quickly; how they might be planned and built; what they could look like; and how we can get them, by a committee of The Association of Building Technicians – 11 people of whom Walter Segal was one. Edited by Colin Penn & Andrew Boyd, foreword by Minister of Health Aneurin Bevan, Paul Elek, London, 1946.

1948
Home and Environment, Leonard Hill, London 1948 (2nd ed 1953). Important book, based on extensive series of articles about house design and layout and town planning which Segal had been publishing regularly, particularly in *The Architect & Building News* and *Building*, from 1942–47.

1949
»For the First Time: A Letter to an Architect about to Spend a Holiday in Switzerland« *A&BN*, 5 August 1949, pp126–128.

1953
»Small Houses« (lead article; written anonymously by Segal for Monica Pidgeon) *AD*, November 1953, p299.

1958
»Planning or Snobbery« (long and important letter on the aesthetic control of planners) *AJ*, 27 November 1958, p771.

1959
»Justice and the architect« (two-page feature) *AJ*, 15 January 1959, pp112–3.

1960
»Banham's Book« (Review of *Theory and Design in the First Machine Age* by P R Banham) *AJ*, 28 July 1960, p141.

1964
»The Use of Land in relation to Building Height, Coverage and Housing Density« Lecture at the Architectural Association, 16 January 1964; *A A Journal*, March 1964, pp 253–258.
»Trend, countertrend or counterdrift« (important, long letter) *AJ*, 29 January 1964, p226–7.

1965
»The Unknown Giant« *AJ*, 8 September 1965, pp 523–530. An extended and thoughtful Le Corbusier obituary.

1966
»Changing Trends in Site Layout« *Arena (The Journal of the Architectural Association)*, (No 900, Vol 8) March 1966, pp231–235. See also *AD*, September 1967, p430–1.

1969
»Architecture; The Assertive and the Unobtrusive« *A&BN*, 25 September 1969, pp24–33. First coherent publication of Segal's notion of Modern architecture.

1970
»Mart Stam: Pioneer and Perfectionist« (text from the exhibition) *RIBAJ*, July 1970, p317.
»A Future Tense for Architects« *A&BN*, 17 September 1970, pp32–9.

1971

»Beyond Utility: Architecture and the Id« *The Architect* (successor to *A&BN*), March 1971, pp38–42, partly republished *AJ*, 4 May 1988. Important statement of Segal's ideas on meaning in architecture.

»Low Cost Housing and User Participation«, paper given as one of four in a One-Day Symposium, University of Edinburgh, April 1971, published in: *Architecture and Social Sciences*, (booklet), edited by Dr P G Raman, University of Edinburgh, pp 96–132. Segal's clearest statement of the development of his approach.

1972

»About Taut« *The Architectural Review (AR),* January 1972,pp25 –7. Feature article based on review of Bruno Taut by Kurt Junghanns.

»Case study of three houses in use« *AJ*, 26 January 1972, pp209 – 15. Segal's own clinical examination of his three houses in Tasker Road NW3 in use.

»Things Can't Go On As They Are« *The Architect*, February 1972, pp46–7.

»The Neo-Purist School of Architecture« *Architectural Design*, June 1972, pp344–5.

1973

»Scharoun« *AR,* February 1973, pp98–102. Essay on expressionism, Taut and others, based on occasion of Scharoun's death.

»Looking Back to the Architecture of the 1920s before and after« First Bannister Fletcher Lecture, Bartlett School of Architecture and Planning, University College London, 16 May 1973 (edited version »Into the Twenties«, *AR*, January 1974, pp 31–8).

»Meaning and Non-Meaning in Architecture« Second Bannister Fletcher Lecture, Bartlett School of Architecture and Planning, University College London, 23 May 1973 (largely based on »Beyond Utility«, *The Architect*, March 1971; see above).

»Practice« Third Bannister Fletcher Lecture, Bartlett School of Architecture and Planning, University College London, 6 June 1973.

»Home... Sweet Home?« *RIBA Journal*, October 1973, pp 477–480. History and critique of 85 years of London housing.

1974

»Less is More« *AJ*, 20 February 1974, p372. Extended review of *Small is Beautiful* by E Fritz Schumacher.

»Are we now going to tackle the real problem?« *AJ*, 20 November 1974, pp1199–20. One in a series of comments solicited by John McKean (AJ Housing Editor) on UK Government housing policy.

1976

»A Time for Change« *AJ*, 1 September 1976, pp392–4. Criticism of building regulations and planning structure.

»Architecture in Wood« in *The International Book of Wood*, ed Martyn Bramwell, Mitchell Beazley Publishers Ltd., London, 1976, pp51–81. This essay was extensively edited and rewritten for publication.

1977

»The housing crisis in Western Europe: Britain – Assessment and Options« in *Architecture for People*, ed. Byron Mikellides Studic Vista, London, 1977.

»Timber Framed Housing« *RIBAJournal*, July 1977, pp284–95. First autobiographical lecture (in »Architect's Approach to Architecture« series) at RIBA May 1977, (edited tape transcription).

1982

»Walter Segal; View from a Lifetime« *RIBA Transactions*, 1, Vol 1, No 1, 1981–2, pp7–14. Second lecture given at RIBA on 2 March 1982, (edited tape transcription).

1983

»Learning from the Self-Builders«, »Tape/Slide« package, Pidgeon Audio Visual, PAV9/8301, 1983. Unscripted »lecture«, describing slides; the introduction recorded last.

In *AR* in 1972, Mark Girouard asked WS: »What journals have you contributed to?« And WS replied: »amongst others – *Oeuvres* (Geneva), *Jardins* (Paris), *VivendasD (Barcelona), AJ, AR, RIBA Journal, Arena, Architect & Building News, Building, Byggmaesteren* (Stockholm), *Wasmuths Monatshefte für Baukunst, Architectural Design.«

There exist fairly complete bibliographies of Segal in two versions, one compiled by the author and the other by Peter Blundell Jones, each with the other's assistance and help from Segal's own periodical collection (owned by John Segal), Charlotte Ellis, Philip Christou and others.

Selected Writing about Walter Segal
Literatur über Walter Segal

1930
»Wettbewerbe – 2 Preis Walter Segal« (plan and perspective of Segal project illustrated), *Bauwelt*, Vol 29, 1930, p922. First publication of a Segal project.

1931
»Walter Segal, K H Beyerling, Adolf Bloch« *Bauwelt*, Vol 9, 26 February 1931, p20. Description of student competition entry for timber-framed house. Reprinted with a statement by Segal (and colleagues) »Why we use Heraklith« in *Heraklith Rundschau*, September 1931, p7.

1935
»Villa à Palma di Majorca«, Julius POSENER. *L'Architecture d'Aujourdhui*, January 1935, pp 42–3. First publication of a Segal building. »Maison d'été à Ascona: architecte Walter Segal«, *Maisons et Jardins*, (cutting undated, probably 1935?), pp26–7.

1937
»Commended School Design« *The Architects' Journal* (AJ), 25 March 1937, pp541–4. Article on the News Chronicle School Competition with commended design of Walter & Eva Segal.
La Casa Piccola: *The Architectural Review* (AR), Coronation Number, May 1937, pp 215–6; in F R S YORKE, *The Modern House*, 3rd ed, The Architectural Press, London, 1937 pp118–9;in E H B BOULTON, *Timber buildings in the country*, Country Life, 1938, pp528–9.

1939
Timber furniture by Segal *Country Life*, 11 March 1939 pLXIV and another article in *Wood*, May 1939, p223.

1941
»Family shelters« *Building*, January 1941, pp12–16. Segal's design for air-raid shelters.

1942
»Hostel for factory workers« Architect Bertram Carter, chief assistant Walter Segal *Architect & Building News (A&BN)*, v170, 15 May 1942, pp96–9 *and A&BN*, 22 May 1942, pp111–5.

1943
»A Hot-Water Supply Unit« (Architect: Walter Segal) *Architectural Design & Construction*, November 1943, pp231–5.
Other Segal heating unit designs: *Coke and Smokeless Fuel Age*, February 1946, pp26–30, and *A&BN*, 19 April 1946, pp41–4.

1947
»A Housing Scheme for Winchester. Architects Alexander Kurtz and Walter Segal« *Building*, October 1947, pp315–319.
»Proposed House at Mill Hill« *A&BN*, 28 November 1947, pp180–6.

1948
Reviews of *Home and Environment* included:

C S MARDALL, *Journal of the Royal Institute of British Architects (RiBAJournal)*, June 1948, p368; Hugh CASSON, *Sunday Times*, 18 July 1948; and numerous others including *Architectural Record* (USA), *l'Architecture d'Aujourd'hui* (France) and *Bauen und Wohnen* (Germany).

»Flats in Leigham Court Road by Walter Segal« *Architectural Design* (AD), December 1948, p260. See also *AD* July 1952, p194.

1949
»The house that Jack can build today: architect Walter Segal«, *Good Housekeeping*, January 1949 pp22–4, first in a series on new houses (which Segal edited).
»The Abse House: architect Walter Segal« *A&BN*, 22 April 1949, pp355–8.
»Designing the small house: four examples by Walter Segal« *Building*, August 1949, pp284–8.

1953
»House at Hendon, London by Walter Segal« (A house at West Heath Gardens illustrated) *AD*, November 1953, pp315–6. See also *Architecture and Building*, January 1954, p22–6.

1955
Article on Tretol building, clearly based on text by Segal and incorporating direct quotations *A&BN*, April 1955, pp132–7.
Working Detail: Tretol staircase *AJ*, 4 August 1955. (Republished in *Architects' Working Details*, Vol 4, The Architectural Press, London, 1957, (Segal details on pp 22–3, 42–3).

1958
Working Details: Flats in London *AJ*, 3 July 1958, p34; another in *AJ*, 17 July 1958, p106; Factory in London *AJ*, 31 July 1958; House in Surrey *AJ*, 27 November 1958, p771 (republished in *Architects' Working Details*, Vol 7, The Architectural Press, London, 1960, (Segal details on pp 34–5, 54–5, 94–5).
»Factory and Warehouse in Ramsgate Street, London« *AJ*, 2 October 1958, pp 493–500. Building Study on premier pickle factory.

1960
Working Detail: Fireplace at St Anne's Close, Highgate, N6 *AJ*, 28 July 1960, pp 159–60.

1962
Working Details Revisited: Tretol *AJ*, 28 November 1962, p1237 and *AJ*, 5 December 1962, p1300; spiral staircase at Compayne Gardens *AJ*, 29 May 1963, pp1135–6; Ovington Square SW1 windows *AJ*, 4 August 1965, pp267–9 stairs *AJ*, 18 August 1965, p370; Pickle factory

AJ, 25 August 1965, p432 and *AJ*, 4 September 1965, p494; own fireplace *AJ*, 22 September 1965, p670.

1966
Temporary house: *AJ*, 26 January 1966, p252, fully illustrated *AJ*, 23 March 1966, pp763–769; also *The Financial Times*, Wednesday, 23 March 1966, p10; *Ideal Home*, September 1966, pp38–41 and January 1967 pp91–5.

1968
»A Man on His Own« by Kate WHARTON *A&BN*, 23 October 1968, pp20–7; reprinted (partially) *AJ*, 4 May 1988. First article about Segal's practice, largely in his own words, based on interview.

1969
Halstead timber house: *The Daily Telegraph Magazine*, 28 March 1969, pp46–50; *Essex County Standard and Halstead Gazette*, 15 August 1969, p3.
»Wirtschaftlich Bauen, Ein Gespräch mit Walter Segal«, Wilhelm KAINRATH *Deutsche Bauzeitung*, 1 October 1969, pp755–8. Interview with Segal.

1970
»Low Cost Housing Scope«, Greg CONDLITTLE, Architecture East Midlands, May/June 1970, p27. Article on Segal's system) »Segal's Significance«, Wilhelm KAINRATH *AJ*, 30 September 1970, pp769–80 and *ac revue* January 1971. General article on timber houses up to 1970.

1974
»Preforming prefabs«, report on Segal lecture at AA, *Ghost Dance Times* (AA paper, edited Martin Pawley), 29 November 1974, p3.

1975
»A Certain Basic Satisfaction in Building a Shelter for Oneself«, John McKEAN *AJ*, 3 September 1975, pp458–61. Report on first self-built Segal house.
»Das Segal-Konzept: Häuser aus der Baustoffhandlung«, H. D. GULICHER *Baumeister* 11, November 1975, pp956–7. General article on Segal's building system.
»Segal self-build in Lewisham« *AJ*, 5 November 1975, p934. The genesis of the Lewisham housing. And »Do-It-Yourself Housing« John McKEAN Building Design (BD), 30 April 1976, pp14–5. Discussion of implications of Local Authority self-build at Lewisham.
»Housing; Lightweight Timber System« Richard MacCORMAC and David LEA *AJ*, 26 November 1975, pp1121–38.

1976
»Walter Segal, pioneer: his early life and the Modern Movement«, and »Walter Segal, pioneer: transition to self-build timber housing«, John McKEAN *BD*, 20 February 1976, pp10–11 and *BD*, 27 February 1976, pp18–9. Extended profile of Segal and his work.
»The Segal System« (Essay in special issue of *AD* »Whatever Happened to the Systems Approach?«) John McKEAN *AD*, Vol XLV1, May 1976, pp267–303. Discussion of Segal's ›method‹, illustrating his documentation.

1977
»Segal's Other Buildings« DS (Deyan SUDJIC) *AJ*, 20 April 1977, pp716–8.

1978
»The Anarchy of Planning« John McKEAN *BD*, 17 March 1978, pp14–15, Segal's bureaucratic problems in building housing for Solon in Bromley, London.

»Eine Baumethode für Selbstbau« M. Gross and J. Bareib, *F + I – bau*, April 1978, pp13–9. See also same author's article on student adaptation of Segal: *Deutsche Bauzeitung*, January 1979, pp43–5.
»Groot-tegen kleinschadligheid de discussie gaat door« Myriam DARU, Plan (Holland), No 6, 1978, p47–54.

1979
Self-build housing groups and co-operatives, Peter STEAD, Anglo-German foundation, 1979, mentions Segal frequently and the Toothill project by Segal in detail.

1980
»The Green House« *AJ*, 27 August 1980, pp404–5. Report on house at Sharnbrook, Bedfordshire.
»Do-it-yourself vernacular« Charlotte ELLIS *AJ*, 17 December 1980, pp1185–1205. Study of Lewisham first self-build scheme. (See also leading article p1173).
»Walter Segal« Peter BLUNDELL JONES in *MacMillan encyclopaedia of architects*, 1980. General essay.

1981
»Walter Segal – Biografie eines Aussenseiters (oder: Warum vernünftige Häuser für vernünftige Leute zu bauen heute ein subversives Unternehmen ist)« Wilhelm KAINRATH in ›Für eine andere Architektur‹, edited by Michael Andritzky et al., Frankfurt, 1981, pp98–107
»People and Ideas: a small triumph of Lewisham«, Colin WARD, *Town & Country Planning*, March 1981, pp92–3. On Lewisham, also: *ac revue*, April 1981, pp32–3; *BD*, 25 September 1981; Open House (Eindhoven) Vol 7, No 3, pp14–23. *The Guardian*, 2 April 1982; »Open Space« BBC-TV programme made by the Lewisham self-builders, transmitted BBC2 April 1982; *AMC* (France), No 9, October 1985.

1982
»Segal's First Half-Century«, Charlotte ELLIS, *AJ*, 7 April 1982.

1983
»Student Self-build in Stuttgart«, Peter BLUNDELL JONES, *AJ*, 27 February 1983, pp32–50 and *Spazio e Societa* March 1984, pp106–125. Work inspired by Segal.

1984
»Self-Build Selection« , Charlotte ELLIS, *AJ*, 25 January 1984, pp37–39. Second Lewisham project.
»Building revisit: Walter Segal's House«, Martin PAWLEY, *AJ*, 20 June 1984, pp35–8. The ›temporary‹ house after 20 years.
Obituaries: Alan BLANC, *BD*, 1 November 1985, p2; »Walter's Widsom« leader in *AJ*, 6 November 1985, p27; Charlotte ELLIS (and others) *AJ*, 6 November 1985; Colin WARD, *New Society* 15 November 1985, p295; Peter RICH, *A3 TIMES*, Polytechnic of North London, No. 7, Autumn 1985, p1.
Extended appreciation written when Segal died: John McKEAN »Semi Preziosi di Buon Senso« (Italian version) and »Lift High the Roof beams, Carpenters!« (English version) *Spazio e Societa (Space and Society)*, Anno 9, No. 34, Giuno/June 1986, pp18–26 reprinted as »Walter Segal: The man and the myth« *BD*, 20 May 1988 pp15–19.

1986
»The Segal Method« Jon BROOME *AJ*, 5 November 1986, pp31–68. Special issue of *AJ* clearly explaining and illustrating the building method.

1987

»Das Segal-System: Diskurs zu einer Methode«, John McKEAN *Archithese* (CH), January 1987, pp22–32.

»Homes Fit for Humans«, Charlotte ELLIS, *AR*, March 1987, pp82–5. Background article on Segal's ideas, and appraisal of second Lewisham project.

1988

»Walter Segal 1907–1985«, *AJ*, 4 May 1988. Special issue to coincide with Segal memorial exhibition at Royal Festival Hall, London, May–June 1988. Reprint of various Segal »astragal« gossip notes; Essays by Julius POSENER and Peter BLUNDELL JONES; article on ›temporary‹ house and Tasker Road; reprint of two Segal articles; »Segal's Legacy« by Graham VICKERS; also good outline biography, list of buildings and writing.

»Working with Walter«, Alan BLANC, *BD*, 20 May 1988, pp20–3. St Anne's Close and life as Segal's assistant around 1950.

»Learning the Lewisham way«, *AJ*, 18 May 1988, p87, Nicholas TAYLOR. Institutional client's-eye view of Lewisham.

»The go-between«, Colin WARD, paper at *Walter Segal Symposium*, Purcell Room, South Bank, 7 June 1988. Allusive image of Segal as a ›go-between‹.

There are also unpublished student dissertations on Segal, including those by: D. Duval (Portsmouth 1972); M. G. Penny (1973); H. C. Smyth (Newcastle 1973); M. P. Goodhart (May 1978); A. Riley (South Bank 1980); H. Y. Keun (Thames 1983); B. Wills (Newcastle 1984); P. Sutton (PNL 1987).

Credits
Bildnachweis

We thank John A. Segal for his kind permission to use all original material, both published and archival, by Walter Segal.

64 of the 200 illustrations used herein are from Walter Segal's archive which has been researched, ordered and catalogued by Philip Christou and Florian Beigel since Segal's death. Their research was done for the purpose of presenting Segal's ideas in the exhibition which was first seen at the Royal Festival Hall in London in May 1988. (See page 188). We thank Florian Beigel and Philip Christou for their cooperation in providing these illustrations.

The quotation from Paul Valery's »Eupalinos, or the Architect« is from pp20–21 of W. McC. Stewart's translation, published in 1932 by Oxford University Press and quoted with their permission.

We credit all illustrations as accurately as possible below. The abbreviation »SA« refers to the *Segal archive*; »AP« to *The Architectural Press Ltd*, London; »H&E« to Segal's book *Home and Environment*, Leonard Hill, London, 1948.

In der Reihe »Architektur im Zusammenhang« bereits erschienen:

Rudolf Schilling
Der Hang und Zwang zum Einfachen
Ausblick auf eine andere Wohnarchitektur
208 Seiten, 125 Abbildungen
ISBN 3-7643-1640-3

Dieter Bachmann, Gerardo Zanetti
Architektur des Aufbegehrens
Bauen im Tessin
192 Seiten, 100 Abbildungen
ISBN 3-7643-1731-0

Erich Bramhas
Der Wiener Gemeindebau
Vom Karl Marx-Hof zum Hundertwasserhaus
196 Seiten, 204 Abbildungen
ISBN 3-7643-1797-3

Rudolf Schilling
Rückbau und Wiedergutmachung
Was tun mit dem gebauten Kram?
192 Seiten, 195 Abbildungen
ISBN 3-7643-1912-7